CONFRON THE NATION'S WATER PROBLEMS

The Role of Research

ittee on Assessment of Water Resources Research

Water Science and Technology Board
Division on Earth and Life Studies

NATIONAL RESEARCH COUNCIL
OF THE NATIONAL ACADEMIES

THE NATIONAL ACADEMIES PRESS
Washington, D.C.
www.nap.edu

THE NATIONAL ACADEMIES PRESS 500 Fifth Street, N.W. Washington, DC 20001

NOTICE: The project that is the subject of this report was approved by the Governing Board of the National Research Council, whose members are drawn from the councils of the National Academy of Sciences, the National Academy of Engineering, and the Institute of Medicine. The members of the committee responsible for the report were chosen for their special competences and with regard for appropriate balance.

This study was supported by Cooperative Agreement No. 02HQAG0086 between the National Academy of Sciences and the U.S. Geological Survey. Any opinions, findings, conclusions, or recommendations expressed in this publication are those of the author(s) and do not necessarily reflect the views of the organizations or agencies that provided support for the project.

International Standard Book Number 0-309-09258-2 (Book)
International Standard Book Number 0-309-53335-X (PDF)

Library of Congress Catalog Card Number 2004112038

Confronting the Nation's Water Problems: The Role of Research is available from National Academies Press, 500 Fifth Street, N.W., Lockbox 285, Washington, DC 20055; (800) 624-6242 or (202) 334-3313 (in the Washington metropolitan area); Internet, http://www.nap.edu

Printed in the United States of America

THE NATIONAL ACADEMIES

Advisers to the Nation on Science, Engineering, and Medicine

The **National Academy of Sciences** is a private, nonprofit, self-perpetuating society of distinguished scholars engaged in scientific and engineering research, dedicated to the furtherance of science and technology and to their use for the general welfare. Upon the authority of the charter granted to it by the Congress in 1863, the Academy has a mandate that requires it to advise the federal government on scientific and technical matters. Dr. Bruce M. Alberts is president of the National Academy of Sciences.

The **National Academy of Engineering** was established in 1964, under the charter of the National Academy of Sciences, as a parallel organization of outstanding engineers. It is autonomous in its administration and in the selection of its members, sharing with the National Academy of Sciences the responsibility for advising the federal government. The National Academy of Engineering also sponsors engineering programs aimed at meeting national needs, encourages education and research, and recognizes the superior achievements of engineers. Dr. Wm. A. Wulf is president of the National Academy of Engineering.

The **Institute of Medicine** was established in 1970 by the National Academy of Sciences to secure the services of eminent members of appropriate professions in the examination of policy matters pertaining to the health of the public. The Institute acts under the responsibility given to the National Academy of Sciences by its congressional charter to be an adviser to the federal government and, upon its own initiative, to identify issues of medical care, research, and education. Dr. Harvey V. Fineberg is president of the Institute of Medicine.

The **National Research Council** was organized by the National Academy of Sciences in 1916 to associate the broad community of science and technology with the Academy's purposes of furthering knowledge and advising the federal government. Functioning in accordance with general policies determined by the Academy, the Council has become the principal operating agency of both the National Academy of Sciences and the National Academy of Engineering in providing services to the government, the public, and the scientific and engineering communities. The Council is administered jointly by both Academies and the Institute of Medicine. Dr. Bruce M. Alberts and Dr. Wm. A. Wulf are chair and vice chair, respectively, of the National Research Council.

www.national-academies.org

COMMITTEE ON ASSESSMENT OF WATER RESOURCES RESEARCH

NRC Staff

Preface

At the dawn of the 21st century the United States faces a panoply of water problems that are significantly more numerous, complex, and larger in scope than those of the past. Serious issues about how water resources are to be protected and managed are not confined to one or several regions; they are found nationwide. Increasingly, the science needed to resolve these water issues in workable ways is not available. Thus, for example, efforts to resolve water allocation problems that stem from the need to protect endangered species in the Klamath and Missouri River basins are being constrained by the lack of good scientific information upon which to base policies which will simultaneously protect biodiversity and minimize adverse economic consequences. Efforts to protect and enhance water quality are similarly hamstrung by the absence of scientific information which would allow water managers to respond proactively to both new and existing threats. And, despite the importance of aquatic ecosystems in generating both service and amenity values, our scientific understanding of how such systems function is rudimentary.

This report, which was undertaken at the request of Congress, illuminates the state of the water resources research enterprise in the United States. It is the logical sequel to an earlier volume entitled: *Envisioning the Agenda for Water Resources Research in the Twenty-First Century,* prepared by the Water Science and Technology Board of the National Research Council (NRC). The purpose of the present report is to:

- refine and enhance the findings of the Envisioning report

- examine current and historical patterns and magnitudes of investment in water resources research at the federal level and assess its adequacy
- address the need to better coordinate the nation's water resources research enterprise
- identify institutional options for the improved coordination, prioritization and implementation of research on water resources

Thus, the report discusses the history of federal support of water resources research, it proposes a framework for prioritizing the research agenda, it reports the results of a survey which was intended to describe the characteristics of the current national research effort, and it addresses issues related to the future organization of water resources research at the federal level. Because the NRC committee assembled to conduct the study was charged with examining the water resources research enterprise in the United States, the report does not encompass international water resources research endeavors, including those research efforts funded by American agencies in foreign countries. The committee acknowledges that some future review might usefully take an international perspective.

The committee found that federal investment in water resources research has remained essentially static (in real terms) for more than 30 years. Moreover, water research is accomplished in a highly decentralized fashion with numerous federal agencies setting research agendas independently of each other. Much of that research is focused on operational and near-term problems, with less attention and fewer resources devoted to longer-term, more fundamental research. There are a number of circumstances that suggest *a priori* that the nation's water resources research enterprise is not as well organized and financed as it will need to be if the science required to address the multiplying water problems confronting the United States is to be available.

In developing this report, the committee benefited greatly from the advice and input of a companion group of federal agency and non-governmental organization liaisons established for the purpose of assisting the committee. Individual liaisons are identified in Appendix F. The committee also benefited from discussions held with a group of state representatives at its second meeting in January 2003. The list of state representatives can be found in Appendix D. We thank all those who took time to share with us their perspectives and wisdom about the various issues affecting the water resources research enterprise.

The committee was ably served by the staff of the Water Science and Technology Board and its director, Stephen Parker. Study director Laura Ehlers kept the committee on task and on time and provided her own valuable insights which have improved the report immeasurably. Anita Hall provided the committee with all manner of support in a timely and cheerful way. This report would not have been possible without the help of these people.

This report has been reviewed by individuals chosen for their diverse perspectives and technical expertise, in accordance with the procedures approved by the NRC's Report Review Committee. The purpose of this independent review is to provide candid and critical comments that will assist the authors and the NRC in making the published report as sound as possible and to ensure that the report meets institutional standards of objectivity, evidence, and responsiveness to the study charge. The reviews and draft manuscripts remain confidential to protect the integrity of the deliberative process. We thank the following individuals for their participation in the review of this report: John Boland, Johns Hopkins University; Patrick Brezonik, University of Minnesota; Robert A. Frosch, Harvard University; Gerald E. Galloway, Titan Corporation; Peter H. Gleick, Pacific Institute for Studies in Development, Environment, and Security; Bernard Goldstein, University of Pittsburgh; George M. Hornberger, University of Virginia; Judy L. Meyer, University of Georgia; and James Westcoat, University of Illinois.

Although the reviewers listed above have provided many constructive comments and suggestions, they were not asked to endorse the conclusions or recommendations nor did they see the final draft of the report before its release. The review of this report was overseen by Floyd E. Bloom, The Scripps Research Institute, and Daniel P. Loucks, Cornell University. Appointed by the NRC, they were responsible for making certain that an independent examination of this report was carried out in accordance with institutional procedures and that all review comments were carefully considered. Responsibility for the final content of this report rests entirely with the committee.

Henry Vaux, Jr.
Chair

Contents

Executive Summary

Nothing is more fundamental to life than water. Not only is water a basic need, but adequate safe water underpins the nation's health, economy, security, and ecology. The strategic challenge for the future is to ensure adequate quantity and quality of water to meet human and ecological needs in the face of growing competition among domestic, industrial–commercial, agricultural, and environmental uses. To address water resources problems likely to emerge in the next 10–15 years, decision makers at all levels of government will need to make informed choices among often conflicting and uncertain alternative actions. These choices are best made with the full benefit of research and analysis.

In June 2001, the Water Science and Technology Board of the National Research Council (NRC) published a report that outlined important areas of water resources research that should be addressed over the next decade in order to confront emerging water problems. *Envisioning the Agenda for Water Resources Research in the Twenty-first Century* was intended to draw public attention to the urgency and complexity of future water resource issues facing the United States. The report identified the individual research areas needed to help ensure that the water resources of the United States remain sustainable over the long run, with less emphasis on the ways in which the setting of the water research agenda, the conduct of such research, and the investment allocated to such research should be improved.

Subsequent to release of the *Envisioning* report, Congress requested that a new NRC study be conducted to further illuminate the state of the water resources research enterprise. In particular, the study charge was to (1) refine and enhance the recent findings of the *Envisioning* report, (2) examine current and historical patterns and magnitudes of investment in water resources research at the federal level and generally assess its adequacy, (3) address the need to better coordinate the nation's water resources research enterprise, and (4) identify institutional options for the improved coordination, prioritization, and implementation of research in water resources. The study was carried out by the Committee on Assessment of Water Resources Research, which met five times over the course of 15 months.

The committee was motivated by considering the following central questions about the state of the nation's water resources: (1) will drinking water be safe; (2) will there be sufficient water to both protect environmental values and support future economic growth; (3) can effective water policy be made; (4) will water quality be enhanced and maintained; (5) will our water management systems adapt to climate change? If the answers to even some of the questions above are "no," it would portend a future fraught with complex water resource problems but with limited institutional ability to respond. Knowledge and insight gained from a broad spectrum of natural and social science research on water resources are key to avoiding these undesirable scenarios.

Two realities helped to shape the scope of the study and have illuminated the inherent difficulties in creating a national agenda for water resources research. First, the type and quantity of research that will be needed to address current and future water resources problems are unlikely to be adequate if no action is taken at the federal level. For many reasons (as discussed in Chapter 1), the states and nongovernmental organizations have limited incentives and resources to invest in water resources research. Furthermore, most states are experiencing an increasing number of complex water problems—some of which cross state lines—and they have to respond to important federal mandates. This suggests a more central role for the federal government in producing the necessary research to inform water resources issues. Second, water resources problems do not fall logically or easily within the purview of a single federal agency but, rather, are fragmented among nearly 20 agencies. As water resource problems increase in complexity, even more agencies may become involved. The present state of having uncoordinated and mission-driven water resources research agendas within the federal agencies will have to be changed in order to surmount future water problems.

Chapter 2 of this report analyzes the history of federally funded water resources research in an effort to understand how the research needed to solve tomorrow's problems may compare with the research undertaken in the past, and to illuminate how U.S. support for water resources research in the 20th century has fluctuated in response to important scientific, political, and social movements. Federal support of water-related research developed slowly during much of the 1800s and early 1900s, beginning with federal involvement in the development of rivers for navigation, flood control, and storage of water for irrigation. It was not until the 1950s that Congress committed to supporting a comprehensive program of water resources research. The short-lived commitment peaked during the 1960s when Congress and the executive branch shared a similar view that the federal role in water entailed funding its development for human use while reducing problems of pollution. By the 1970s, growing interest in environmental protection conflicted with water development, which splintered the policy consensus and cast the federal government into more of a regulatory role while deemphasizing its role in promoting economic growth through water resources development.

Administrations of the 1980s and 1990s asserted a more limited federal role in water resources research, believing that research should be closely connected to helping to meet federal agency missions or to addressing problems beyond the scope of the states or the private sector. Congress, on the other hand, generally supported a broader approach to water research, but one that it could actively supervise through the legislative and appropriations process. A consequence of the devolving of responsibility for water resources research back to the states was the neglect of long-term, basic research as opposed to the favoring of applied research that would lead to more immediate results.

Over the last 50 years, the priority elements of a national water resources research agenda have been identified in widely varying ways by many organizations and reports. Many general topics of concern—for example, water-based physical processes, availability of water resources for human use and benefit, and hydrology–ecology relationships—have appeared repeatedly over the decades, while others, such as the impact of climate change and newly discovered waterborne contaminants, are recent topics. The reappearance of some of the same topics over time suggests that the nation's research programs, both individually and collectively, have not responded in an adequate manner and that there is no structure in place to make use of the research agendas generated by various expert groups. Indeed, at the national level there is no coordinated process for considering water resources research needs, for prioritizing them for funding purposes, or for evaluating the effectiveness of research activities.

In the face of the historical inability to mount an effective, broadly conceived national program of water resources research, it is reasonable to ask, "Why bother with yet another comprehensive proposal?" The answer lies in the sheer number of water resource problems (as illustrated in Chapter 1) and the fact that these problems are growing in both number and intensity. To address these problems successfully, the nation must invest not only in applied research but also in fundamental research that will form the basis for applied research a decade hence. A repeat of past efforts will likely lead to enormously adverse and costly outcomes for the status and condition of water resources in almost every region of the United States.

A METHOD FOR SETTING PRIORITIES OF A NATIONAL RESEARCH AGENDA

The solution to water resource problems is necessarily sought in research—inquiry into the basic natural and societal processes that govern the components of a given problem, combined with inquiry into possible methods for solving these problems. In many fields, the formulation of explicit research priorities has a profound effect on the conduct of research and the likelihood of finding solutions to problems.

Water resources research areas were extensively considered in the *Envisioning* report, resulting in a detailed, comprehensive list of 43 research needs, grouped into three categories. The category of *water availability* emphasizes the interrelated nature of water quantity and water quality problems, and it recognizes the increasing pressures on water supply to provide for both human and ecosystem needs. The category of *water use* includes not only research questions about managing human consumptive and nonconsumptive use of water, but also about the use of water by aquatic ecosystems and endangered or threatened species. The third category, *water institutions*, emphasizes the need for research into the economic, social, and institutional forces that shape both the availability and use of water. Interestingly, input from federal and state government representatives gathered during the course of this project confirmed the importance of many of the 43 topics.

Rather than focusing on a topic-by-topic research agenda, this report identifies overarching principles to guide the formulation and conduct of water research. Indeed, statements of research priorities developed by a group of scientists or managers can, depending on the individuals, have a relatively narrow scope. In recent years, the limitations of discipline-based perspectives have become clear, as researchers and managers alike have recognized that water problems relevant to society necessarily integrate across physical, chemical, biological, and social sciences. Furthermore, research priorities should shift as new problems emerge and past problems are mitigated or brought under control through scientifically informed policy and actions. Thus, Chapter 3 provides a mechanism for reviewing, updating, and prioritizing the current water resources research agenda (as expressed in the *Envisioning* report) and subsequent versions of the agenda. This mechanism is much more than a summing up of the priorities of the numerous federal agencies, professional associations, and federal committees. Rather, it consists of six questions or criteria (listed below) that can be used to assess individual research priorities and thus to assemble (and periodically review) a responsive and effective national research agenda.

1. **Is there a federal role in this research area?** This question is important for evaluating the "public good" nature of the water resources research area. A federal role is appropriate in those research areas where the benefits of such research

are widely dispersed and do not accrue only to those who fund the research. Furthermore, it is important to consider whether the research area is being or even can be addressed by institutions other than the federal government.

2. **What is the expected value of this research?** This question addresses the importance attached to successful results, either in terms of direct problem solving or advancement of fundamental knowledge of water resources.

3. **To what extent is the research of national significance?** National significance is greatest for research areas (1) that address issues of large-scale concern (for example, because they encompass a region larger than an individual state), (2) that are driven by federal legislation or mandates, and (3) whose benefits accrue to a broad swath of the public (for example, because they address a problem that is common across the nation). Note that while there is overlap between the first and third criteria, research may have public good properties while not being of national significance, and vice versa.

4. **Does the research fill a gap in knowledge?** If so, it should clearly be of higher priority than research that is duplicative of other efforts. Furthermore, there are several common underlying themes that, given the expected future complexity of water resources research, should be used to evaluate research areas:

- the interdisciplinary nature of the research
- the need for a broad systems context in phrasing research questions and pursuing answers
- the incorporation of uncertainty concepts and measurements into all aspects of research
- how well the research addresses the role of adaptation in human and ecological response to changing water resources

These themes, and their importance in combating emerging water resources problems, are described in detail in Chapter 3.

5. **How well is this research area progressing?** The adequacy of efforts in a given research area can be evaluated with respect to the following:

- current funding levels and funding trends over time
- whether the research area is part of the agenda of one or more federal agencies
- whether prior investments in this type of research have produced results (i.e., the level of success of this type of research in the past and why new efforts are warranted)

6. **How does the research area complement the overall water resources research portfolio?** When applied to federal research and development, the portfolio concept is invoked to mean a mix of fundamental and applied research; of shorter-term and longer-term research; of agency-based, contract, and investigator-driven research; and of research that addresses both national and region-specific problems—with data collection to support all of the above. Indeed, the priority-setting process should be as much dedicated to ensuring an appropriate balance and mix of research efforts as it is to listing specific research topics.

The following conclusions and recommendations are made about the creation and refinement of a national portfolio of water resources research.

The 43 research topics from the *Envisioning* report are the current best statement of research needs, although this list is expected to change as circumstances and knowledge evolve. Water resource issues change continuously, as new knowledge reveals unforeseen problems, as changes in society generate novel problems, and as changing perceptions by the public reveal issues that were previously unimportant. Periodic reviews of and updates to the priority list are needed to ensure that it remains not only current but proactive in directing research toward emerging problems.

An urgent priority for water resources research is the development of a process for regularly reviewing and revising the entire portfolio of research being conducted. The six questions listed above are helpful for assessing both the scope of the entire water resources research enterprise and the nature, urgency, and purview of individual research areas. Addressing these questions should ensure that the vast scope of water resources research carried out by the numerous federal and state agencies, nongovernmental organizations, and academic institutions remains focused and effective.

The research agenda should be balanced with respect to time scale, focus, source of problem statement, and source of expertise. Water resources research ranges from long-term and theoretical studies of basic physical, chemical, and biological processes to studies intended to provide rapid solutions to immediate problems. The water resources research enterprise is best served by developing a mechanism for ensuring that there is an appropriate balance among the different types of research, so that both the problems of today and those that will emerge over the next 10–15 years can be effectively addressed.

The context within which research is designed should explicitly reflect the four themes of interdisciplinarity, broad systems context, uncertainty, and adaptation. The current water resources research enterprise is limited by the agency missions, the often narrow disciplinary perspective of scientists, and the lack of a national perspective on perceived local but widely occurring problems. Research patterned after the four themes articulated above could break down these barriers and promise a more fruitful approach to solving the nation's water resource problems.

STATUS AND EVALUATION OF WATER RESOURCES RESEARCH IN THE UNITED STATES

In order to evaluate the current investment in water resources research, the committee collected budget data and narrative information in the form of a survey from the major federal agencies and significant nonfederal organizations that are conducting water resources research. The format of the survey was similar to an accounting of water resources research that occurred from 1965 to 1975 by the Committee on Water Resources Research of the Federal Council for Science and Technology. This earlier effort entailed annually gathering budget information from all relevant federal agencies in 60 categories of water resources research. In order to support a comparison of the current data with past information, the NRC committee adopted a modified version of the earlier model, using most of the same categories and subcategories of water resources research. In January 2003, the survey was submitted to all of the federal agencies that either perform or fund water resources research and to several nonfederal organizations that had annual expenditures of at least $3 million during one of the fiscal years covered by the survey. See Table 4-1 for a complete list of respondents.

The survey consisted of five questions related to water resources research (see Box 4-1). In the first question, the liaisons were asked to report total expenditures on research in fiscal years 1999, 2000, and 2001 for 11 major categories (and 71 subcategories) of water resources research. (All data collection activities were explicitly excluded from the survey.) The remaining questions were posed to help give the committee a better understanding of current and projected future activities of the agencies, to provide a qualitative understanding of how research performance is measured, and to gauge the agencies' mix of research, in terms of fundamental vs. applied, internal vs. external, and short-term vs. long-term research. Responses to the survey were submitted in written form and orally at the third committee meeting, held April 29–May 1, 2003, in Washington, D.C.; revised survey responses submitted by the liaisons in summer 2003 reflected corrections and responded to specific requests from the committee.

Evaluation of the submitted information included a trends analysis for the total amount of water resources research funding and for the funding of the 11 major categories of water resources research. The total budget for water resources research from 1965 to 2001 and the year 2000 breakdown by federal agency are shown in Figures ES-1 and ES-2. The budget data were also analyzed to determine the extent to which the 43 high-priority research areas in the *Envisioning* report are being addressed. Finally, the committee qualitatively assessed the

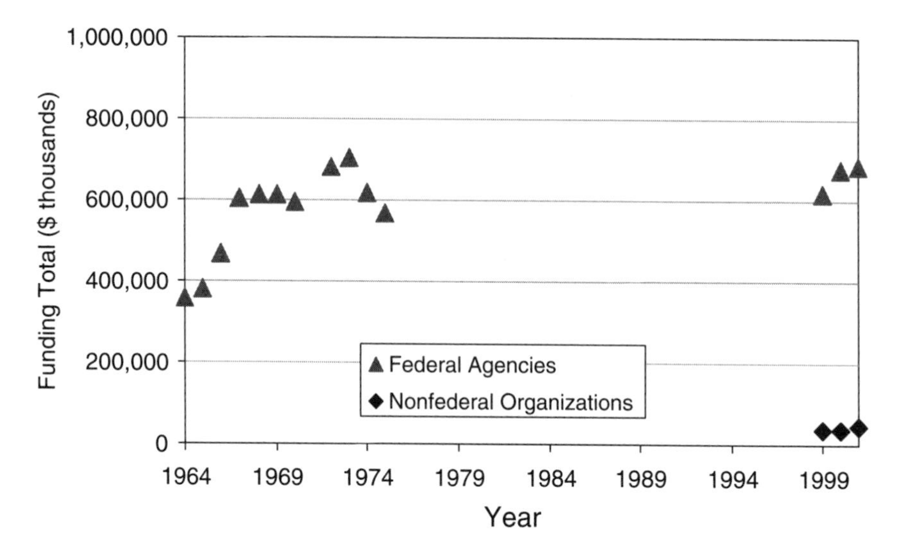

FIGURE ES-1 Total expenditures on water resources research by federal agencies and nonfederal organizations, 1964–2001. Values reported are FY2000 dollars. No survey data are available for years 1976–1988.

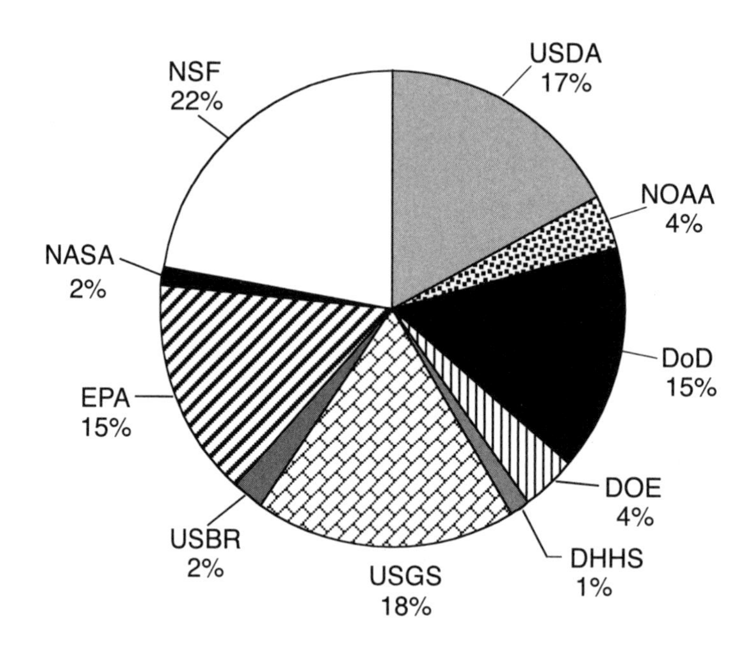

FIGURE ES-2 Agency contributions as a percentage of the total federal funding for water resources research in 2000.

balance of the current national water resources research portfolio (defined as the sum of all agency-sponsored research activities). The following conclusions and recommendations stem directly from these evaluations.

Real levels of total spending for water resources research have remained relatively constant (around $700 million in 2000 dollars) since the mid 1970s. When Category XI (aquatic ecosystems) is subtracted from the total funding, there is a very high likelihood that the funding level has actually declined over the last 30 years. It is almost certain that funds in Categories III (water supply augmentation and conservation), V (water quality management and protection), VI (water resources planning and institutional issues), and VII (resources data) have declined severely since the mid 1970s. All statements about trends are supported by a quantitative uncertainty analysis conducted for each category.

Water resources research funding has not paralleled growth in demographic and economic parameters such as population, gross domestic product (GDP), or budget outlays (unlike research in other fields such as health). Since 1973, the population of the United States has increased by 26 percent, the GDP and federal budget outlays have more than doubled, and federal funding for all research and development has almost doubled, while funding for water resources research has remained stagnant. More specifically, over the last 30 years water resources research funding has decreased from 0.0156 percent to 0.0068 percent of the GDP, while the portion of the federal budget devoted to water resources research has shrunk from 0.08 percent to 0.037 percent. The per capita spending on water resources research has fallen from $3.33 in 1973 to $2.40 in 2001. Given that the pressure on water resources varies more or less directly with population and economic growth, and given sharp and intensifying increases in conflicts over water, a new commitment will have to be made to water resources research if the nation is to be successful in addressing its water and water management problems over the next 10–15 years.

The topical balance of the federal water resources research portfolio has changed since the 1965–1975 period, such that the present balance appears to be inconsistent with current priorities as outlined in Chapter 3. Research on social science topics such as water demand, water law, and other institutional topics, as well as on water supply augmentation and conservation, now garners a significantly smaller proportion of the total water research funding than it did 30 years ago. When the current water resources research enterprise is compared with the list of research priorities noted in the *Envisioning* report, it becomes clear that significant new investment must be made in water use and institutional research topics if the national water agenda is to be addressed adequately. If enhanced funding to support research in these categories is not diverted from other categories (which may also have priority), the total water research budget will have to be enhanced.

The current water resources research portfolio appears heavily weighted in favor of short-term research. This is not surprising in view of the de-emphasis of long-term research in the portfolios of most federal agencies. It is important to emphasize that long-term research forms the foundation for short-term research in the future. A mechanism should be developed to ensure that long-term research accounts for one-third to one-half of the portfolio.

The Office of Management and Budget (OMB) should develop guidance to agencies on reporting water resources research by topical categories. Understanding the full and multiple dimensions of the federal investment in water resources research is critical to making judgments about adequacy. In spite of clearly stated OMB definitions of research, agencies report research activity unevenly and inconsistently. Failure to fully account for all research activity undermines efforts by the administration and Congress to understand the level and distribution of water resources research. This problem could be remedied if OMB required agencies to report all research activity, regardless of budget account, in a consistent manner.

DATA COLLECTION AND MONITORING

Although data collection was excluded from the water resources research survey conducted by the committee, the long-term monitoring of hydrologic systems and the archiving of the resulting data are critical to the water resources research enterprise of the nation. Data are essential for understanding physicochemical and biological processes and, in most cases, provide the basis for predictive modeling. Long-term consistent records of data, which capture the full range of interannual variability, are especially essential to understanding and predicting low-frequency, high-intensity events. Furthermore, federal agencies are instrumental in developing new monitoring approaches, in validating their efficacy through field studies, and in managing nationwide monitoring networks over long periods. The following conclusions and recommendations address the need for investments in basic data collection and monitoring.

Key legacy monitoring systems in areas of streamflow, groundwater, sediment transport, water quality, and water use have been in substantial decline and in some cases have nearly been eliminated. These systems provide data necessary for both research (i.e., advancing fundamental knowledge) and practical applications (e.g., for designing the infrastructure required to cope with hydrologic extremes). Despite repeated calls for protecting and expanding monitoring systems relevant to water resources, these trends continue for a variety of reasons.

The consequences of the present policy of neglect associated with water resources monitoring will not necessarily remain small. New hydrologic problems are emerging that are of continental or near continental proportions. The scale and the complexity of these problems are the main arguments for improvements to the *in situ* data collection networks for surface waters and groundwater and for water demand by sector. It is reasonable to expect that improving the availability of data, as well as improving the types and quality of data collected, should reduce the costs for many water resources projects.

COORDINATION OF WATER RESOURCES RESEARCH

Coordination of the water resources research enterprise is needed to make deliberative judgments about the allocation of funds and scope of research, to minimize duplication where appropriate, to present Congress and the public with a coherent strategy for federal investment, and to facilitate the large-scale multiagency research efforts that will likely be needed to deal with future water problems. Unfortunately, water resources research across the federal enterprise has been largely uncoordinated for the last 30 years, although there have been periodic ad hoc attempts to engage in interagency coordination during that time. The lack of coordination is partly responsible for the topical and operational gaps apparent in the current water resources research portfolio. Thus, although the federal agencies are carrying out their mission-driven research, most of this work focuses on short-term problems, with a limited outlook for crosscutting issues, longer-term problems, and more basic research that often portends future solutions. As a result, it is not clear that the sum of individual agency priorities adds up to a truly comprehensive list of national needs and priorities.

There are few areas of research as broadly distributed across the federal government as water resources research, resulting in few examples of how to effectively coordinate large-scale research programs. Nonetheless, the committee identified those factors that encourage or discourage effective coordination of large-scale research programs after hearing about programs for highway research, agricultural research, earthquake and hazard reduction research, and global change research. These factors helped shed light on an effective model for coordination of water resources research, which relies on some entity performing the following functions:

- doing a regular survey of water resources research using input from federal agency representatives
- advising OMB and Congress on the content and balance of a long-term national water resources research agenda every three to five years

- advising OMB and Congress on the adequacy of mission-driven research budgets of the federal agencies
- advising OMB and Congress on key priorities for fundamental research that could form the core of a competitive grants program
- engaging in vertical coordination with states, industry, and other stakeholders, which would ultimately help refine the agenda-setting process

The first three activities are intended to make sure that there is a national agenda for water resources research, that it reflects the most recent information on emerging issues, and that the water resources missions of the federal agencies are contributing in some way to national agenda items. A competitive grants program (the fourth activity) is proposed as a mechanism for filling critical gaps in the research portfolio, in the event that certain high-priority research areas are not being adequately addressed by the federal agencies and to increase the proportion of long-term research. This program would require new (but modest) funding. Given the topical gaps noted earlier and in Chapter 4, **funding would be needed on the order of $20 million per year for research related to improving the efficiency and effectiveness of water institutions and $50 million per year for research related to challenges and changes in water use.**

Three institutional models that could conceivably carry out the bulleted activities listed above are described in Chapter 6. The first model relies on an existing interagency body—the Subcommittee on Water Availability and Quality administered by the Office of Science and Technology Policy. This coordination option is attractive because arrangements are already in place and agency roles and responsibilities are well defined. However, this approach has yet to demonstrate that it can be an effective forum for looking beyond agency missions to fundamental research needs. The second option involves Congress authorizing a neutral third party to perform the functions above, which would place the outside research and user communities on equal footing with federal agency representatives. The independence from the agencies afforded by this option makes it possible to focus the competitive grants program on longer-term research needs, particularly those falling outside agency missions. A disadvantage is that it may engender resentment from the agencies, and OMB may be reluctant to establish such a formal advisory body. A third option is a hybrid model that would be led by OMB and formally tied to the budget process. For more detailed descriptions the reader is referred to Chapter 6, which comprehensively discusses the three options.

Any one of the three coordination options could be made to work in whole or part. Each has strengths and weaknesses (described in detail in Chapter 6) that would need to be weighed against the benefits and costs that could accrue from moving beyond the status quo. In the end, decision makers will choose the coordination mechanism that meets perceived needs at an acceptable cost in terms of level of effort and funding. It is possible that none of the options is viable in its entirety. However, it may be possible to partially implement an option, which in itself would be an improvement over the status quo. For example, the initiation of a competitive grants program targeted at high-priority but underfunded national priorities in water resources research could occur under any one of the options and in lieu of the other activities listed above.

* * *

Publicly funded research has played a critical role in addressing water resources problems over the last several decades, both for direct problem solving and for achieving a higher level of understanding about water-related phenomena. Research has enabled the nation to increase the productivity of its water resources, and additional research can be expected to increase that productivity even more, which is critical to supporting future population and economic growth. Managing the nation's water resources in more environmentally sensitive and benign ways is more important than ever, given the recognition now afforded to aquatic ecosystems and their environmental services. A course of action marked by the creation and maintenance of a coordinated, comprehensive, and balanced national water resources research agenda, combined with a regular assessment of the water resources research activities sponsored by the federal agencies, represents the nation's best chance for dealing effectively with the many water crises sure to mark the 21st century.

1

Setting the Stage

In the coming decades, no natural resource may prove to be more critical to human health and well-being than water. Yet, there is abundant evidence that the condition of water resources in many parts of the United States and the world is deteriorating. Our institutions appear to have limited capacity to manage water-based habitats to maintain and improve species diversity and provide ecosystem services while concurrently supplying human needs. In some regions of the country, the availability of sufficient water to service growing domestic uses is in doubt, as is the future sufficiency of water to support agriculture in an increasingly competitive and globalizing agricultural economy. Indeed, demands for water resources to support population and economic growth continue to increase, although water supplies to support this growth are fixed in quantity and already are fully allocated in most areas. Renewal and repair of the aging water supply infrastructure, particularly along the eastern seaboard, will require time and hundreds of billions of dollars (GAO, 2002). These are examples of a mounting array of water-related problems that touch virtually every region of the country and for which scientifically sound and economically feasible solutions need to be found.

The future water crisis is unlikely to materialize as a monolithic catastrophe that threatens the livelihoods of millions. Rather it is the growing sum of hundreds, perhaps thousands, of water problems at regional and local scales (and not just in the semiarid West, as interstate conflicts over new water supplies for the metropolitan Washington, D.C., region and Atlanta, Georgia, testify). Indeed, a search of the *New York Times*, Associated Press, and Reuters databases for articles related to "water" or "wetlands" found over 330 articles for 2002 alone, with 29 of the 50 states being the subject of at least one article.[1] Indications of the increasing frequency of significant water-based environmental problems include such events as the recent collapses of ecosystems in the Gulf of Mexico and the Chesapeake Bay. Increases in damages attributable to droughts and floods are evidence of the nation's vulnerability to extreme weather events. The threat of waterborne disease, as exemplified by the 1994 *Cryptosporidium* outbreak in Milwaukee, Wisconsin, and subsequent less dramatic events, is constantly present. Nonetheless, it is difficult to perceive the increasing frequency of these problems because water resources management and research tend to be highly decentralized.

In much the same way that it makes the totality of the nation's water problems difficult to comprehend, decentralization also masks the extent to which scientific information is required to address these problems. Yet, as numerous cases in this report illustrate, making good decisions about water issues requires scientific understanding and, thus, continued investment in water resources research. The growing complexity of water problems only reinforces this need for scientific information in fashioning new and innovative solutions. Unfortunately, although the number, complexity, and severity of water problems are growing, investment in the scientific research needed to develop a better understanding of water resources and the ways in which they are managed has stagnated. Overall investment in research on water and water-related topics has not grown in real terms over the last quarter century, even as the number of relevant research topics has expanded. Much of the current federal and state research agenda tends to focus on short-term problems of an operational nature. Too little of it is focused on the kind of fundamental, integrated, longer-term research that will be required if current and emerging water problems are to be addressed successfully. Furthermore, research agendas are not normally prioritized (from either a regional or national perspective), with the result that there is no assurance that the research being done is focused on the most urgent and important problems. Also, there is no assurance that the ad hoc research agendas that do emerge lead to efficient investment among the research priorities.

ISSUES OF CONCERN IN WATER RESOURCES

The magnitude of water resources problems, and the importance of research in addressing them, are best illustrated by referring to specific examples, a number of which are described below.

---

[1]This search was conducted on the *New York Times* web site (www.nytimes.com) for calendar year 2002 using the words "water" and "wetlands." Article title and summaries were searched to verify that the articles were about current local, regional, or national water resources problems.

Will Drinking Water Be Safe?

Over the past 100 years, investment in water research as well as in water treatment and distribution infrastructure has made the quality of drinking water in the United States among the best in the world. Enormous gains in public health were realized from the virtual elimination of typhoid and cholera, which were once spread through the water supply. Today, the provision of safe and reliable supplies of drinking water is taken for granted in the United States. Nonetheless, new chemical contaminants and biological agents continue to emerge and threaten the safety of water supplies. For example, the inorganic chemical perchlorate was discovered in drinking water wells in northern California in 1997 (AWWARF, 1997), having found its way into groundwater from manufacturing processes (for rocket fuels, munitions, and fireworks) and inadequate disposal practices. Perchlorate is now known to interfere with thyroid hormone production and is a suspected human carcinogen, and it has been shown to affect the drinking water supplies of more than 12 million consumers in at least 14 states (Renner, 1998). Other contaminants await discovery and, like perchlorate, will be added to the U.S. Environmental Protection Agency's (EPA) drinking water Contaminant Candidate List, which already contains 60 chemical and microbial species awaiting regulatory determinations (EPA, 1998). In addition to the periodic appearance of new contaminants that result from inadvertent lapses in the handling and disposal of chemicals, the potential for intentional contamination of drinking water supplies now represents a real and continuing threat.

Appropriate treatment of drinking water supplies often requires trade-offs that are sometimes not well understood scientifically. For example, membrane-based treatment technologies, such as reverse osmosis, that remove contaminants from drinking water ultimately concentrate the contaminants in another medium, the disposal of which would be of environmental concern and could pose a threat to the health and safety of workers who must handle the material. This type of trade-off must be clearly characterized and understood if the most reliable and cost-effective methods of treating the nation's drinking water are to be developed. Much additional research will be needed to (1) identify biological and chemical constituents that could threaten water supplies and (2) identify methods of treatment to remove existing and future contaminants without creating toxic hazards or additional problems with environmental contamination. Technologies that deal with multiple contaminants while minimizing associated health and environmental effects are especially needed.

Will There Be Sufficient Water to Support Both the Environment and Future Economic Growth?

The semiarid states of the American West and Southwest are the fastest-growing states in the nation and will require new supplies of urban water. Yet, the waters of these states, which are naturally in short supply, are almost all fully allocated among environmental, urban, and agricultural uses. And the existing mechanisms for reallocating water away from current uses are not well developed and are frequently ineffective. The problem lies with the fact that there are no well-functioning institutions that allow people in these regions to live with an essentially fixed supply of water. The result is frequently paralyzing political conflict, as the examples below demonstrate.

In 2000, the U.S. Fish and Wildlife Service (FWS) concluded that the current flow regime of the Missouri River jeopardized at least three rare and endangered species—the pallid sturgeon, least tern, and piping plover. FWS recommended modifications in the criteria used by the U.S. Army Corps of Engineers to guide dam operations on the Missouri River, which would entail both regular increases in spring flows and reduced summer flows (lower summer flows are intended to support nesting and foraging habitat for least terns and the piping plover, as well as nursery habitat for the pallid sturgeon and other fishes). Because such a change in flow regimes would materially impinge upon the navigability of the Missouri and make waterway transportation difficult or impossible during the harvest season, some stakeholders have challenged the scientific validity of the finding. The resulting political impasse over how to manage the flows of the Missouri means that the riverine ecosystem continues to be degraded with the accompanying threat to the three species, while agricultural and waterway transportation interests have no guarantee that their positions will ultimately prevail. While acknowledging that there are some gaps in the science underlying the FWS biological opinion, the National Research Council (NRC) concluded that changes are necessary in the flow regime, which should be adaptively managed to reduce scientific uncertainties (NRC, 2002).

The story is very much the same in the Klamath Basin of southern Oregon and northern California. As widely reported, in 2002 agricultural producers engaged in civil disobedience when their irrigation water was cut off so that river flows and lake levels could be enhanced to support several endangered fish species. These growers, who suffered real economic damages, have criticized the biological opinion that led to the shut-off as being scientifically inadequate. And, in a situation reminiscent of the Missouri River confrontation, an NRC report raised doubts about the science underlying the Klamath opinion (NRC, 2003).

Some have argued that these cases are simply the tip of the iceberg—that the Missouri and the Klamath are forerunners to dozens of such conflicts about to emerge—and that there is a lack of adequate science to support balanced resolutions of these conflicts. The real fear is that economic growth will be restrained without any tangible improvement in the quality of the environment. Nor are these cases limited to the western United States and to surface water. Glennon (2002) describes examples from all parts of the United States in which the overwithdrawals of groundwater have led to the drying up of streams and rivers, leading to cascading social conflicts, economic hardships, and ecological degradation. In many cases, this is exacerbated by the legal institutions that regulate groundwater, which tend to ignore the physical realities of hydrology and the tremendous increase in the scientific understanding of hydrology since the pertinent regulations were first enunciated. With respect to problems in the East, the city of Atlanta, Georgia, is engaged in a struggle with Alabama and Florida to acquire adequate water to support growth in its metropolitan area. The waters of the Potomac River are the focus of a dispute between Virginia and Maryland over allocative doctrines that date back to the 1700s—a dispute that ultimately had to be resolved by the U.S. Supreme Court. Although some allocative disputes may be resolved in the courts, in all regions of the country better water science is needed to make improvements in the efficiency with which water is used, and to help ensure that water is allocated and reallocated in a balanced way that acknowledges the need to support economic growth, agricultural productivity, and environmental protection.

Can Effective Water Policy Be Made?

There is evidence that many of our federal water policies are either ineffective or only partially effective. A good example is the "no net loss" policy for wetlands initially adopted in 1989. At that time, there was much concern over the fate of the nation's wetlands, as it was variously estimated that more than half of the nation's wetlands had been converted to other uses (Dahl, 1990). It is widely recognized that wetlands are among the most biologically productive environments and that they provide important environmental services such as flood protection and water quality maintenance (NRC, 1995). In 1989, then President George H. W. Bush promulgated a policy of "no net loss" of wetland area and function, which continues to be the policy of the federal government.

In implementing this policy it was recognized that economic development and agricultural activity inevitably result in the degradation or destruction of wetlands. Hence the government required that damages to the nation's wetlands be mitigated. A recent review of this policy (NRC, 2001a) concluded that despite a requirement that more than one acre of wetland be restored or created for each acre lost, only 69 percent of the acreage required was actually restored or created, and the type of wetland resulting from the mitigation action was often different from and of lower ecological value than the wetland that was lost. Further, up to 90 percent of the mitigation efforts were not monitored, and there was full compliance with only 55 percent of the wetland permits issued. The report concluded that the failure of mitigation policies to protect wetlands results from shortcomings in policy making and implementing institutions as well as from inadequacies in our current understanding of restoration ecology. The nation's ability to improve compensatory mitigation will depend upon the way in which mitigation practices are administered, monitored, and recorded, which will require a considerably enhanced scientific understanding of the structure and processes of wetlands (NRC, 2001a).

The management of wetlands is not the only water policy to be hampered by a lack of scientific information. Policies governing (1) the treatment of drinking water supplies, (2) the use of water in agriculture, (3) the maintenance and preservation of aquatic habitats and species diversity, (4) the treatment and reuse of wastewater, and (5) the management of floods and droughts will all require additional scientific information if they are to be effective.

Can Water Quality Be Maintained and Enhanced?

During the 1970s and 1980s, the nation made good progress in improving surface water quality. Through a strategic combination of permitting requirements and financial support for the construction of municipal wastewater treatment facilities, dramatic improvements were realized in the quality of the nation's surface waters. Yet, the failure to deal with nonpoint source pollutants has come to represent an important omission in national water quality management. Beginning in the 1980s efforts were made to implement a Total Maximum Daily Load (TMDL) program that had been originally authorized by the Clean Water Act of 1972. Rules were devised and states were required to implement TMDL programs for impaired waters, which would lead (among other things) to the control of nonpoint source discharges. This has proved to be difficult, and successes are limited.

In 2000, the U.S. General Accounting Office (GAO) reported that the primary impediment to the successful implementation of state TMDL programs was the lack of high-quality data and information to make fundamental decisions (GAO, 2000), such as decisions about which waterbodies are in violation of water quality standards, about the extent to which nonpoint pollutants contribute to the problem in question, and about how TMDLs should be calculated for waterbodies that are in violation of standards. Only five of the 50 states claimed to have the tools and information needed even to assess all of their waters (GAO, 2000).

A subsequent study by the NRC identified major gaps in the knowledge required to make TMDL programs effective (NRC, 2001b). Creating this knowledge will require research leading to the development of more refined statistical tools, of watershed and water quality models, and of innovative bioassessment techniques. A much-needed updating of the antiquated 1992 TMDL Rule has now been stalled, in part because the tools and techniques required for full and effective implementation of the TMDL program are only now under development. The larger picture is that as new chemicals are manufactured, as the legacy of chemical plumes in the soil from early agricultural and industrial activities is visited upon groundwater, and as conversion of wildlands and other relatively undisturbed lands continues, the need for more and better science to support our understanding of contaminant behavior and treatment and control technologies will increase.

Will Our Water Management Systems Adapt to Climate Change?

Existing data show that there have been unprecedented changes in climate in the postindustrial era (Jones et al., 1999; Karl and Trenberth, 2003). There are, moreover, legitimate concerns about the nature and pace of future climate changes and the implications of those changes for water resources (Gleick et al., 2000; IPCC, 2001). Although there are many remaining uncertainties about the scope, intensity, and timing of climate change in the coming decades, there is scientific agreement about the projected occurrence of change for a wide range of water-related issues. For example, there is a high degree of confidence that rising temperatures will alter snowfall and snowmelt dynamics in the western United States, affecting the timing and magnitude of both winter and spring runoff and forcing changes in reservoir operations. Similarly, higher sea levels along the coasts will increase salinity contamination in coastal freshwater aquifers and alter coastal marshes and wetlands (Gleick et al., 2000). Furthermore, research is beginning to suggest that there will be changes in various types of extreme climatic events (Meehl et al., 2000). Thus, for example, higher frequencies of extreme warm days, lower frequencies of extreme cold days, a decrease in diurnal temperature ranges associated with higher nighttime temperatures, increased precipitation intensity and extremes, and midcontinental summer drying have been widely predicted. Reliable prediction of the frequency of occurrence of extreme weather events is particularly important because of the implications of such events for water availability, food supply systems, and plant and human health. The midwestern floods of 1993 and the drought of 1988 provide examples of the vulnerability of agricultural and urban ecosystems to such events (Rosenzweig et al., 2001).

Embedded within the overall global climate variability, the El Niño-Southern Oscillation (ENSO) phenomenon is an important quasi-periodic cycle of the equatorial Pacific Ocean that affects the earth's climate and has been shown to be the precursor of floods and droughts in various regions of the world (Ropelewski and Halpert, 1996; Harrison and Larkin, 1998). Climate modeling simulations show that in the future, there is likely to be an increased frequency of ENSO-related events that will be of greater severity, significantly affecting regional renewable water supplies (Timmermann et al., 1999). In view of such predictions, consideration of such phenomena in regional water resources planning studies is now warranted. Unfortunately, although substantial resources have been expended for basic climate change research (e.g., NSTB, 2002), little funding has been provided to translate these research findings into new and improved methods for water resources planning and management—an area that has been identified as the weakest element of climate change integrated assessments (NAST, 2001).

New research is needed to help cope with the large uncertainty associated with how climate change will affect water resources and to develop new institutional approaches for water resources and risk management under intensified climate variability. Indeed, professional water groups and agencies have now begun to ask for research into how climate change will affect water systems (AWWA, 1997).

* * *

Will drinking water be safe? Will there be sufficient water to protect environmental values and support future economic growth? Can effective water policy be made? Will water quality be enhanced and maintained? Will our water management systems adapt to climate change? All of these questions address issues that bear on the overarching question to which this report is addressed: **In the future, will there be adequate water to meet the needs of competing users?** What if the answer to this question or to some of the derivative questions discussed above is "no"? It would portend a very difficult future—one in which the water supplies are not always available or are available only in very limited quantities; water quality continues to deteriorate, reducing available supplies; our ability to devise policies to manage water resources effectively is severely constrained; and we struggle, unsuccessfully, to adapt to climate change. A vibrant and robust research program alone will not be sufficient to prevent all of these scenarios, but knowledge and insight gained from a broad spectrum of natural and social science research on water resources is society's best hope for success.

The type and quantity of research that will be needed to address current and future water resources problems are unlikely to be available if no action is taken at the federal level. Although the states are frequently vested with the responsibility to resolve many water resource problems and to respond to federal mandates such as the Clean Water Act, 13 state representatives who met with the committee were unanimous in stating that the increasing number of water problems as well as their increasing complexity are rapidly eclipsing the states' ability to resolve those problems. State governments typically do not have substantial research capacity in the water resource topics, nor do they have the funding to support research on the scale needed to solve their collective problems (for example, if the problems are basinwide in nature, and thus transcend the boundary of a given state). Moreover, individual states may not have an incentive to conduct water resources research, especially if the results would be broadly applicable to more than just the sponsoring state. That is, individual states have a disincentive to conduct water resources research because it has the characteristics of a pure public good (as discussed in the succeeding section). Nongovernmental organizations will similarly underinvest in water resources research due to both weak incentives and the lack of financial resources. Taken together, this suggests that the federal government will need to fund or produce the necessary research to address future water crises, if it is to be produced at all.

Nor do water resources problems fall logically and easily within the purview of a single federal agency. Indeed, as discussed in this report, federal responsibilities for water resources management and research are fragmented among nearly 20 agencies. As water resource problems increase in complexity, even more agencies may become involved. At the present time, the uncoordinated and mission-driven water resources research agendas of these agencies (see Chapters 4 and 6) are inadequate to meet the challenges that lie ahead. In the absence of a new and strong commitment at the federal level to generate additional knowledge of all kinds related to water resources, the future may be characterized by never-ending strife and frustration over our inability to surmount water problems growing in number and complexity.

WHY PUBLICLY SUPPORTED RESEARCH?

It is sometimes asked why the federal government should support research on water resources instead of leaving this activity to the private sector. The answer lies with the fact that the results of much water resources research, particularly basic research, have the characteristics of a public good. That is, once the research is concluded, the results should be freely available to many or all, irrespective of whether the recipients directly pay for them. Those who produce research with public good characteristics are unable to capture all of the returns to that research because the results are not patentable or licensable. Indeed, the private sector typically underinvests or fails to invest at all in the production of public goods because it cannot capture or "appropriate" all of the returns from the investment. The problem of the lack of appropriability is especially pertinent to water resources, since water is a publicly held resource. Although private firms and individuals may enjoy the right to use water, they rarely have title to the corpus or body of the resource. Lack of appropriability combined with public ownership of the resource makes the justification for public support of water resources research compelling.

John Wesley Powell, one of the earliest and most distinguished water scientists in the United States, expressed this concept forthrightly as follows:

> Possession of property is exclusive; possession of knowledge is not exclusive; for the knowledge which one man has may also be the possession of another. The learning of one man does not subtract from the learning of another, as if there were a limited quantity of unknown truth. Intellectual activity does not compete with other intellectual activity for exclusive possession of truth; scholarship breeds scholarship, wisdom breeds wisdom, discovery breeds discovery. Property may be divided into exclusive ownership for utilization and preservation, but knowledge is utilized and preserved by multiple ownership. That which

> one man gains by discovery is the gain of other men. And these multiple gains become invested capital, the interest on which is all paid to every owner, and the revenue of new discovery is boundless. It may be wrong to take another man's purse, but it is always right to take another man's knowledge, and it is the highest virtue to promote another man's investigation. The laws of political economy do not belong to the economics of science and intellectual progress. While ownership of property precludes other ownership of the same, ownership of knowledge promotes other ownership of the same, and when research is properly organized every man's work is an aid to every other man's. (Dupree, 1940.)

In the aftermath of World War II, Vannevar Bush wrote a thorough justification for a strong governmental role in supporting research in the scientific community (Bush, 1950). He argued that the responsibilities for promoting new scientific knowledge and for developing scientific talent were properly the concern of the federal government because these activities vitally affect the nation's health, prosperity, and national security. He noted that the benefits of research were widespread and often appeared many years after the research was done.

More recently the National Science Board has endorsed the concepts originally set forth by Bush (National Science Board, 1997). The board noted the emergence of a "global technological marketplace" and the increasing need for knowledge and information to contend with and manage the "modification of natural and social environments that is occurring" on larger scales and at increasingly rapid rates. These trends make the case for governmental support of research even more compelling than it was during the 1940s. The board concluded that changes in circumstances and national priorities do not negate the potential benefits from government-supported research. It is also significant that the board singles out environmental management and "green manufacturing" as areas with public good characteristics for which Bush's original case is particularly cogent today.

There are numerous examples of government-funded research on water resources that has led to significant payoffs for the nation or for distinct regions of the nation. This research falls into two broad categories: (1) that done to facilitate and enhance the solving of water and water management problems and (2) that done to develop the scientific knowledge necessary to undergird mandated regulatory programs.

Problem-Solving Research

Examples of problem-solving research include studies that have facilitated the management of salts in irrigation, research that has permitted more accurate and long-term prediction of weather, and research on the possibilities and arrangements that promote the voluntary transfer of water. These examples are discussed in more detail below.

Managing Salt in Irrigation

When water is applied to crops in irrigated agriculture, it is ultimately transpired by the plant or evaporated from the soil surface, leaving behind salts in the root zone. If salt concentrations are allowed to build up in the root zones, they reduce plant productivity and ultimately result in sterilization of the soil. The failure to manage salt build-up in the root zone is thought to have led to the destruction of many civilizations including the ancient civilization of Mesopotamia. Research conducted primarily by experts at the U.S. Department of Agriculture and to a lesser extent at the nation's universities led to the development of modern techniques for managing salinity in irrigated agriculture; without this research, much of the irrigated land in the semiarid parts of the country would not be productive (ASCE, 1990).

The fundamental technique of salinity management is to apply water (in excess of the crop water requirement) in quantities sufficient to allow salts to be leached below the root zone. The quantities of water needed vary by crop type, as some crops are relatively sensitive to salt and others relatively insensitive (Maas and Hoffman, 1977). However, in many instances the application and leaching of this excess water may cause water tables to rise and may promote the water-logging of soils. Research has revealed that adequate drainage must be part of the overall strategy for managing salt balances. In recent decades, research has focused on the management of drainage waters to maintain and enhance water quality (NRC, 1989). All of this has benefited western growers, many of whom would not have been able to farm on a sustainable basis without the results of these federal research efforts. Needless to say, the entire country has benefited through the provision of affordable food.

Facilitating Voluntary Transfers of Water

The development of water resources in the arid western states focused initially on irrigation, which, given the enormous demands for water to grow crops, consumed the lion's share of available supplies. Under western water law, irrigators as first users hold rights that are fully protected from the claims of more junior users. However, as the western states have grown, demand for water has changed such that portions of the water supply are being shifted from irrigation to new uses. One important strategy to accomplish this is to use voluntary, market-based transactions.

Water rights in western states are regarded as property rights, a key attribute of which is their transferability from one owner to another by sale, lease, or devise. However, historically there had been very few transfers of water rights in western states that have involved actual changes of water use, especially changes from irrigation to other uses. The National Water Commission first identified this problem in the 1970s, after which research that focused on better understanding market-like transactions in water flourished (Meyers and Posner, 1971; Hartman and Seastone, 1970; Johnson et al., 1981; Vaux and Howitt, 1986; Saliba and Bush, 1987). With federal funding, academics from six western states examined water-transfer experiences over a recent 20-year period (Natural Resources Law Center, 1990), and an NRC committee examined issues related to market-like transfers of water in the West (NRC, 1992).

This body of research identified impediments to the transfer of water rights, including the interrelated nature of water use, expensive state review processes not designed to facilitate transfers, and cultural and social factors, and it also investigated how to lessen or eliminate these impediments. A result of the research has been a building up of support for the careful use of water transfers in meeting changing water demands. Water rights transfers are now part of the standard array of options available for solving problems of water scarcity in the West. Transfers allowed Californians to respond effectively to the drought of 1987–1993, which saved millions of dollars. A recently completed accord that includes transfers from the Imperial Irrigation District to San Diego appears to have brought a peaceful resolution to a serious conflict among the states using Colorado River water. Most of this research was supported by the federal government and has resulted in significant benefits to urban dwellers in many parts of the West.

Predicting El Niño

Fundamental research on the nature, manifestation, and impacts of the El Niño phenomenon has been supported by the federal government since the 1970s (e.g., Philander, 1990). Such research efforts received great impetus with the influx of new data from the Tropical Atmosphere Ocean Array of moored buoys installed in the equatorial Pacific by 1994. This research and the availability of new data resulted in the formulation of reliable predictive models prior to the occurrence of the significant El Niño of 1997–1998, an event that developed very rapidly in the first half of 1997 to become one of the highest magnitude El Niño events in the last 50 years (second only to the 1982–1983 El Niño). Although the west coast suffers substantial flood damage when significant El Niño events occur (e.g., Cayan and Webb, 1992), when the 1997–1998 El Niño occurred as predicted, the affected region was well prepared (by adjusting crop planting and fertilizing schedules, sandbagging, altering reservoir release patterns from normal use patterns, accelerating plans to repair and improve structures, etc.). Actual damage was substantially lower than that anticipated had there been no prediction. The National Oceanic and Atmospheric Administration (NOAA) reports that California's emergency management agencies and the Federal Emergency Management Agency spent an estimated $165 million preparing for adverse weather effects prior to the 1997–1998 El Niño, and the actual losses for the event in California were estimated to be $1.1 billion (NOAA, 2002). The losses from the 1982–1983 El Niño were much greater ($2.2 billion). A significant portion of the difference in losses is attributed to increased preparedness resulting from the 1997–1998 El Niño forecast.

Research in Support of Regulation

Examples of federally funded research in support of regulation include studies that broaden our understanding of the causes of eutrophication in inland waters, studies of mercury deposition, and studies to determine the sources of nitrogen loading in the Chesapeake Bay watershed. These are examples of research that have made the regulatory process more effective and fair to those who are subject to it, as discussed below.

Understanding the Causes of Eutrophication

Eutrophication—the result of excessive inputs of nutrients (nitrogen or phosphorus) to fresh and coastal marine waters—is one of the major causes of water quality degradation. The many problems caused by excess nutrient inputs include accelerated growth of phytoplankton both in the water column and in the benthos; the dominance of toxic, bloom-forming algal species; decreases in water transparency; the development of hypoxic and anoxic conditions in the bottom waters and sediments, with concomitant mortality of fish and invertebrates; the production of unpleasant odors and tastes; and interference with filtration of drinking water. Extensive research carried out across a broad range of scales (laboratory-based assays to experimental manipulation of whole lakes) has been necessary to resolve these issues (for a review, see Smith, 1998). This has resulted, among other things, in a range of specific, quantitative assays for distinguishing nitrogen vs. phosphorus limitation in waterbodies and these nutrients' role in controlling phytoplankton production and species composition. A long history of research, beginning with the work of Thienemann (1918) and Naumann (1919) and later expanded by limnologists Golterman (1975), Fruh et al. (1966), Vollenweider (1976), and Schindler (1977) has demonstrated the effect of external nutrient inputs on concentrations of nutrients in lake water and the resulting impairment of waters, allowing regulatory limits to be established. Finally, federally funded research on the role of phosphorus in causing eutrophication led to the adoption of ordinances during the 1970s by some states and cities that banned the use of phosphate-based laundry detergents; these laws have became so widespread that the industry voluntarily eliminated phosphates from all laundry detergents in order to be able to market their products nationally.

Understanding the Risk of Methylmercury

On January 30, 2004, EPA proposed standards of performance for mercury emissions from electric utility steam generating units, the largest single source of mercury emissions. This regulatory proposal, including the controversial cap-and-trade approach, seeks to reduce mercury emissions from coal-fired utility units (EPA, 2004). This proposed rule follows an improved understanding of the risks of mercury exposure for fetuses, infants, and young children obtained through federally funded research.

Methylmercury in humans causes neurological damage that affects memory, attention, and language skills, especially during formative developmental stages. Human exposure to mercury follows a complicated pathway that begins when inorganic mercury from industrial air emissions precipitates over waterbodies. There, it is bacterially transformed into organic forms such as methylmercury, which is far more toxic than inorganic mercury (EPA, 1997). Methylmercury is taken up from the water column by fish, and humans are then exposed following consumption of those fish.

Federally funded research has tackled numerous components of this problem, including studies to determine the level and type of mercury emission sources; studies of mercury fate, transport, and transformation in the water column and on land; studies of human and animal exposure routes for mercury; and risk management studies. A conclusion emanating from some of this research was that there is a plausible link between anthropogenic mercury emissions and methylmercury in fish. Furthermore, a quantitative health risk assessment of methylmercury based on fish consumption surveys estimated that up to 3 percent of women of child-bearing age eat sufficient amounts of fish to put their fetuses at risk from methylmercury exposure (EPA, 1997). These studies have cumulatively led to the naming of mercury as one of the fifteen "pollutants of concern" in the Great Waters[2] (EPA, 2000). The current regulatory proposal reflects this increased understanding of the nature and extent of mercury toxicity and the role of airborne mercury in contaminating water resources.

Nitrogen Loadings in the Chesapeake Bay

In 1983, the first Chesapeake Bay Agreement was adopted, establishing a Chesapeake Bay Program and an executive council to lead restoration efforts within this important estuary. A central element of the program was the setting of targets and timetables for the reduction of phosphorus and nitrogen loading into the Chesapeake Bay. In 1987, the executive council of the program called for a reduction of 40 percent from 1985 levels of "controllable" loads of phosphorus and nitrogen to be achieved by 2000 (Sims and Coale, 2002). These ambitious goals were predicated on the notion that loadings were entering the bay predomi-

[2]The Great Waters are the Great Lakes, Lake Champlain, the Chesapeake Bay, and specific coastal waters designated through the National Estuary Program and the National Estuarine Research Reserve System.

nantly through runoff and groundwater flow. Although the phosphorus target was realistic and achievable by 2000, the nitrogen target was problematic because of the difficulty in reducing fertilizer use, because of suburban development, and because of other diffuse nonpoint discharges into the bay from member states as well as from states further upstream that are not part of the program.

By the early 1990s, federally funded research conducted on airborne pollutant transport led to the estimate that about 25 percent of the nitrogen loadings to the bay originated from long-range transport of emissions from midwestern power plants, motor vehicles, and agricultural operations. These research findings had several important implications for management of the bay. First, they suggested that control of nitrogen loadings to the bay should include an air quality control component that reaches well beyond the existing bounds of the Chesapeake Bay and its catchment area. Second, if the airborne nitrogen could not be controlled, then it would be necessary to impose even more stringent requirements on the known and controllable waterborne sources.

* * *

These examples speak to the role that publicly funded research has played in addressing water resources problems over the last several decades, both for direct problem solving and to achieve a higher level of understanding about water-related phenomena. They illustrate that water problems tend to be regional in nature, but that the benefits of water research tend to be widespread and accrue broadly and to many different user groups. Although not stressed in the above examples, research has allowed the nation to increase the productivity of its water resources, such that today an acre-foot of water yields, on average, more value than it did 50 or 100 years ago.[3] Additional research can be expected to increase that productivity even more, which is critical to supporting future population and economic growth. Finally, research that permits the nation to manage its water resources in more environmentally sensitive and benign ways is more important than ever, given the recognition and value now afforded to aquatic ecosystems and their environmental services.

ENVISIONING THE AGENDA FOR WATER RESOURCES RESEARCH

Recognizing a compelling need for rethinking the national water research agenda in light of emerging as well as persistent problems, the Water Science and Technology Board of the National Research Council in 2001 wrote *Envisioning*

---

[3]This can be documented, both nationally and regionally, by dividing water use by GDP, adjusted for inflation.

the Agenda for Water Resources Research in the Twenty-first Century (NRC, 2001c). The *Envisioning* report was intended to (1) draw attention of the public and broad groups of stakeholders to the urgency and complexity of the water resource issues facing the United States in the 21st century, (2) identify knowledge and corresponding research areas that need emphasis both now and over the long term, and (3) identify ways in which the setting of the water research agenda, the conduct of such research, and the investment allocated to such research should be improved in the near future. The broad goal of the report was to identify the research needed to help ensure that the water resources of the United States remain sustainable over the long run. The report was organized around three broad categories (water availability, water use, and water institutions), and it identified 43 critical water research needs. Conclusions pertinent to the present report include the following:

- the challenge of solving the nation's water problems will require a renewed national research commitment, which will include changes in the way research agendas and priorities are established
- water quality and water quantity need to be thought of in an integrated fashion, and research priorities should be developed in an integrated fashion
- relatively more attention must be given to water-related research in the social sciences and to research focused on the development of innovative institutions than has been the case in the past
- research on environmental water needs has emerged as an important player and should remain a major part of the research agenda.

As discussed below, the present report expands and elaborates upon on this earlier effort.

STATEMENT OF TASK AND REPORT ROAD MAP

The purposes of this report are to (1) refine and enhance the recent findings of the *Envisioning* report, (2) examine current and historical patterns and magnitudes of investment in water resources research at the federal level, and generally assess the adequacy of this investment, (3) address the need to better coordinate the nation's water resources research enterprise, and (4) identify institutional options for the improved coordination, prioritization, and implementation of research in water resources. The study was carried out by the Committee on Assessment of Water Resources Research. The committee has sought to identify overarching principles that will guide the formulation and conduct of water research rather than focusing exclusively on developing a topic-by-topic research agenda. Chapter 2 presents and analyzes the complex history of federally funded water resources research in an effort to understand how the research needed to solve tomorrow's problems may compare with the research undertaken in the past. It is also instructive in illuminating how U.S. support for water resources research in the 20th century has fluctuated in response to important scientific, political, and social movements. Chapter 3 revisits the 43 research areas outlined in the *Envisioning* report, but with the intent of drawing out overarching themes that should govern how research endeavors are organized. It also describes a process for periodically updating the national water resources research agenda. Chapter 4 describes the committee's methodology for collecting budget data and narrative information from the major federal agencies and from significant nonfederal organizations that are conducting water resources research. Within this chapter, these data are analyzed, and conclusions about the nation's investment in water resources research are made. The importance of data collection to the overall water resources research enterprise is the subject of Chapter 5. The report concludes in Chapter 6 with more detailed alternatives for organizing and coordinating the water resources research enterprise than were presented in the *Envisioning* report.

REFERENCES

American Society of Civil Engineers (ASCE). 1990. Agricultural salinity and management. K. K. Tanji (ed.). Water Quality Technical Committee of the Irrigation and Drainage Division of the American Society of Civil Engineers. New York: ASCE.

American Water Works Association Research Foundation (AWWARF). 1997. Report of the Perchlorate Research Issue Group Workshop, Ontario, CA. Denver, CO: AWWARF.

Bush, V. 1950. Science–The Endless Frontier (40th Anniversary Edition). Washington, DC: The National Science Foundation.

Cayan, D. R., and R. H. Webb. 1992. El Niño/Southern oscillation and streamflow in the western United States. Pg. 29–68 *In* El Niño: Historical and Paleoclimatic Aspects of the Southern Oscillation. H. F. Diaz and V. Markgraf (eds.). Cambridge, UK: Cambridge University Press.

Dahl, T. E. 1990. Wetlands losses in the United States 1780s to 1980s. Washington, DC: U.S. Fish and Wildlife Service.

Dupree, A. H. 1940. Science in the Federal Government: A History of Policies and Activities to 1940. Cambridge, MA: Belknap Press of Harvard University Press.

Environmental Protection Agency (EPA). 1997. Mercury Study Report to Congress (Volumes I-VIII). EPA-452/R-97-003 through EPA-452/R-97-010. Washington, DC: EPA Office of Air Quality Planning and Standards and Office of Research and Development.

Environmental Protection Agency (EPA). 1998. Announcement of the drinking water contaminant candidate list. Federal Register 63(40) March 2, 1998.

Environmental Protection Agency (EPA). 2000. Deposition of Air Pollutants to the Great Waters: Third Report to Congress. EPA-453/R-00-005. Washington, DC: EPA Office of Air Quality Planning and Standards and Office of Research and Development.

Environmental Protection Agency (EPA). 2004. Proposed national emission standards for hazardous air pollutants; and, in the alternative, proposed standards of performance for new and existing stationary sources: electric utility steam generating units. Federal Register 69(20) January 30, 2004.

Fruh, G. E., K. M. Stewart, G. F. Lee, G. F. Rohlich, and G. A. Rohlich. 1966. Measurement of eutrophication and trends. Jour. Water Pollut. Cont. Fed. 38:1237–1258.

General Accounting Office (GAO). 2000. Water Quality—Key EPA and State Decisions Limited by Inconsistent and Incomplete Data. GAO/RCED-00-54. Washington, DC: GAO.

General Accounting Office (GAO). 2002. Water Infrastructure: Information on Financing, Capital Planning, and Privatization. GAO 02-764. Washington, DC: GAO.

Gleick, P. H. et al. 2000. Water: The Potential Consequences of Climate Variability and Change for the Water Resources of the United States. The Report of the Water Sector Assessment Team of the National Assessment of the Potential Consequences of Climate Variability and Change for the U.S. Global Change Research Program. Oakland, CA: Pacific Institute for Studies in Development, Environment, and Security.

Glennon, R. 2002. Water Follies: Groundwater Pumping and the Fate of America's Fresh Waters. Washington, DC: Island Press.

Golterman, H. L. 1975. Physiological Limnology: An Approach to the Physiology of Lake Ecosystems. New York: Elsevier Scientific Publishing Co.

Harrison, D. E., and N. K. Larkin. 1998. El Niño-Southern Oscillation sea surface temperature and wind anomalies 1964–1993. Reviews in Geophysics 37(2):353–399.

Hartman, L., and D. Seastone. 1970. Water Transfers: Economic Efficiency and Alternative Institutions. Baltimore, MD: Johns Hopkins University Press.

Intergovernmental Panel for Climate Change (IPCC). 2001. Climate Change 2001: Impacts, Adaptation and Vulnerability. Contribution to the Third Assessment Report to the IPCC. http://www.grida.no/climate/ipcc_tar/wg2/index.htm

Johnson, R., M. Gisser, and L. G. Werner. 1981. The definition of surface water right and transferability. Journal of Law and Economics 273:272–279.

Jones, P. D., M. New, D. E. Parker, S. Martin, and I. G. Rigor. 1999. Surface air temperature and its changes over the past 150 years. Reviews in Geophysics 36(3):353–399.

Karl, T. R., and K. E. Trenberth. 2003. Modern global climate change. Science 302:1719–1723.

Maas, E. V., and G. J. Hoffman. 1977. Crop salt tolerance—current assessment. Journal of the Irrigation and Drainage Division ASCE 103(IR2):115–134.

Meehl, G. A., F. Zwiers, J. Evans, T. Knutson, L. Mearns, and P. Whetton. 2000. Trends in extreme weather and climate events: issues related to modeling extremes in projections of future climate change. Bulletin of the American Meteorological Society 81(3):427–436.

Meyers, C., and R. Posner. 1971. Market Transfers in Water Rights: Toward an Improved Market in Water Resources. National Water Commission Legal Study No. 4. Arlington, VA: National Water Commission.

National Assessment Synthesis Team (NAST). 2001. Climate Change Impacts on the United States: The Potential Consequences of Climate Variability and Change. Washington, DC: U.S. Global Change Research Program. http://www.usgcrp.gov/usgcrp/Library/nationalassessment-/foundation.htm.

National Oceanic and Atmospheric Administration (NOAA). 2002. The economic implications of an El Niño. NOAA Magazine On Line http://www.noaanews.noaa.gov/-magazine/stories/mag24.htm.

National Research Council (NRC). 1989. Irrigation Induced Water Quality Problems. Washington, DC: National Academy Press.

National Research Council (NRC). 1992. Water Transfers in the West: Efficiency, Equity and the Environment. Washington, DC: National Academy Press.

National Research Council (NRC). 1995. Wetlands: Characteristics and Boundaries. Washington, DC: National Academy Press.

National Research Council (NRC). 2001a. Compensating for Wetland Losses Under the Clean Water Act. Washington, DC: National Academy Press.

National Research Council (NRC). 2001b. Assessing the TMDL Approach to Water Quality Management. Washington, DC: National Academy Press.

National Research Council (NRC). 2001c. Envisioning the Agenda for Water Resources Research in the Twenty-first Century. Washington, DC: National Academy Press.

National Research Council (NRC). 2002. The Missouri Ecosystem: Exploring the Prospects for Recovery. Washington, DC: The National Academies Press.
National Research Council (NRC). 2003. Endangered and Threatened Fishes in the Klamath River Basin: Causes of Decline and Strategies for Recovery. Washington, DC: The National Academies Press.
National Science Board. 1997. Government Funding of Scientific Research: A Working Paper of the National Science Board. NSB 97-186. Washington, DC: National Science Board.
National Science and Technology Board (NSTB). 2002. Our Changing Planet. The FY 2002 U.S. Global Change Research Program. A Report by the Subcommittee on Global Change Research, Committee on Environmental, and Natural Resources of the NSTB. A Supplement to the President's Fiscal Year 2002 Budget. Washington, DC: NSTB.
Natural Resources Law Center. 1990. The Water Transfer Process as a Management Option for Meeting Changing Water Demands. Boulder, CO: Natural Resources Law Center.
Naumann, E. 1919. Några synpunkter angående limnoplanktons ekologi med särskild hänsyn till fytoplankton. Svensk Botanisk Tidskrift 13:129–163.
Philander, S. G. 1990. El Nino, La Nina, and the Southern Oscillation. San Diego, CA: Academic Press. 293 pp.
Renner, R. 1998. Perchlorate-tainted wells spur government action. Environmental Science and Technology 32(9):210A.
Ropelewski, C. F., and M. S. Halpert. 1996. Quantifying Southern Oscillation precipitation relationships. Journal of Climate 9:1043–1059.
Rosenzweig, C., A. Iglesias, X. B. Yang, P. R. Epstein, and E. Chivian. 2001. Climate change and extreme weather events: implications for food production, plant diseases and pests. Global Change and Human Health 2(2):90–104.
Saliba, B. C., and D. Bush. 1987. Water Markets in Theory and Practice: Market Transfer, Water Values and Public Policy. Boulder, CO: Westview Press.
Schindler, D. W. 1977. Evolution of phosphorus limitation in lakes. Science 195:260–262.
Sims, J. T., and F. J. Coale. 2002. Solutions to nutrient management problems in the Chesapeake Bay Watershed, USA. Pg. 345–371 *In* Agriculture, Hydrology, and Water Quality. P. M. Haygarth and S. C. Jarvis (eds.). Wallingford, UK: CABI Publishing.
Smith, V. H. 1998. Cultural eutrophication of inland, estuarine and coastal waters. Pg. 7–49 *In* Successes, Limitations, and Frontiers in Ecosystem Science. M. L. Pace and P. M. Groffman (eds.). New York: Springer-Verlag.
Thienemann, A. 1918. Untersuchungen über die Beziehungen zwischen dem Sauerstoffgehalt des Wassers und der Zusammensetzung der Fauna in norddeutschen Seen. Arch. Hydrobiol. 12:1–65.
Timmermann, A., J. Oberhuber, A. Bacher, M. Esch, M. Latif, and E. Roeckner. 1999. Increased El Niño frequency in a climate model forced by future greenhouse warming. Nature 398: 694–697.
Vaux, H. J., Jr., and R. E. Howitt. 1986. Managing water scarcity: an evaluation of interregional transfers. Water Resources Research (20):785–792.
Vollenweider, R. A. 1976. Advances in defining critical loading levels for phosphorus in lake eutrophication. Mem. Inst. Ital. Idrobiol. 33:53–83.

2

The Evolving Federal Role in Support of Water Resources Research

The federal role in helping to ensure that the nation's water resources meet public needs has changed dramatically over the years. Sustained congressional funding to support water resources research started in the 1950s, expanded considerably in the 1960s, and has remained essentially static since then. The research agenda supported by this funding has changed in response to shifting national priorities. This chapter provides a broad overview of the evolving federal role concerning water, starting at the time of European settlement. It then turns to a more detailed discussion of the federal role in water resources research and of efforts over the last 50 years to organize and coordinate federally supported water resources research.

NATIONAL INTERESTS IN WATER

Over the last 200 years, water resources in the United States have undergone a profound transformation. Initially considered as a means of transportation and navigation (and managed as such), water resources for much of the early 20th century were developed primarily as water sources for agriculture and later for industrial and municipal use. The most recent era of water resources management has seen a blossoming of efforts to protect waterbodies from both quality and quantity degradation brought on by such development.

Support of Commerce and Settlement

In this nation's early years, the development of water resources for transportation and other purposes was left largely to private initiative. Congress regarded its powers as narrowly confined to those explicitly conferred upon it by the U.S. Constitution. The earliest federal role in water development and management began in the 1820s when the U.S. Army Corps of Engineers (Corps) was authorized to undertake work to improve the navigability of the nation's coastal and inland waterways. This was made possible by a Supreme Court ruling in the case of *Gibbons* v. *Ogden* (9 Wheat. 1, 197 1824), which declared that Congress had the authority to regulate navigation on interstate rivers under the terms of the interstate commerce clause of the U.S. Constitution. That authority was subsequently broadened by other rulings regarding navigable rivers and their tributaries (Maass, 1951; Hill, 1957). Thereafter, Congress made direct appropriations to the Corps for specific river and harbor improvements. One rationale for authorizing the Corps to be involved in domestic civil works was to provide work for officers during peacetime so as to maintain their engineering proficiency. Navigation improvements continued to be the norm until after the Civil War, when Congress expanded the role and authority of the Corps to include flood control (Holmes, 1972).

During the latter half of the 19th century, the Corps' authority was expanded in two distinct ways. First, the navigation activities that had previously been restricted to maintaining depth in natural channels by clearing debris were expanded to include the construction of dams and other structures on navigable waters. Simultaneously, the Corps also received authority to regulate the disposal of refuse as well as of dredge and fill materials, also for the purpose of protecting navigation. In this way, the Congress could ensure that the activities of the states and of private parties would not interfere with navigation. Second, the Corps was given responsibility for flood control on the lower Mississippi River, which on a regular basis was visited with devastating floods that adversely affected navigation and the development prospects of the region.

To summarize, the federal interests in water resources during the 19th century were limited to matters related to navigation. The expansion of federal authorities into other areas would not occur until the 20th century, although important events in land acquisition and development that set the stage for this occurred in the 19th century (as discussed below).

The period from 1775 to around 1850 has been characterized as the Era of Acquisition in the United States. During this period, the nation acquired most of the lands that would ultimately define the extent of the country on the North American continent, including the Louisiana Purchase in 1803, the annexation of Texas in 1845, and the acquisition of the areas in the southwestern United States, including California, Arizona, and Nevada in 1848 through the Mexican Cession. With the Gadsden Purchase in 1853 and the acquisition of Alaska in 1867, the shape of the United States in North America looked very much the way it looks today.

Beginning around 1850, the Congress made significant efforts to turn much of this land over to private ownership via the Homestead Act of 1862, as large holdings of public lands were seen as inconsistent with democratic ideals. Furthermore, the concept of "Manifest Destiny" dictated early settlement as a means of consolidating western land expanses and the nation's new borders. As this transition occurred, some well-documented abuses on private holdings began to surface, despite Congress's having recently written many of the nation's first grazing, mining, and timber laws. For example, some forested lands were cut-over indiscriminantly, grazing lands were subjected to heavy and nonsustainable grazing pressures, and mining laws were widely abused. This led to federal action between 1880 and 1900 in which some of the remaining timbered lands were reserved to the national forest system and the national park system. It is interesting to note that in creating the national forest system, the Congress made clear that a primary purpose for establishing and managing forest reserves was to "secure favorable conditions of water flows" as a reliable supply for downstream users (16 USCA §475).

In the latter decades of the 19th century, there was concern over the relative absence of settlement and development of broad expanses of land in the arid and semiarid West. Thus, in 1888 the Congress appropriated funds to the U.S. Geological Survey (USGS) to undertake an "irrigation survey" under the leadership of John Wesley Powell.[1] Powell's report, *The Arid Lands of the United States*, underscored the lack of rainfall in the region and the unsuitability of its lands for agriculture without supplemental water through irrigation. Although ultimately rejected by the Congress, Powell's report has become the single most definitive work describing the circumstances of the arid and semiarid West and recommending policies for the development of this region (Pisani, 1992).

The Development Era

The presidency of Theodore Roosevelt between 1901 and 1908 brought with it a new attitude about the federal role respecting natural resources, including water. In place of unfettered private development, Roosevelt promoted the need for governmentally supervised natural resources development to ensure their fullest possible use. This view was prompted to a large extent by the significant acreage of federal land that had been reserved and required management. With Roosevelt's enthusiastic support, Congress passed the Reclamation Act of 1902, committing the federal government to a major role in the development of water resources for use in irrigation in the arid western states.

Roosevelt supported an expanded role for the federal government in the comprehensive development of rivers for economic uses. The report of his Inland

[1]For an excellent account of Powell's life, see Worster (2001). For a more comprehensive treatment of the integration of science into the federal government including the work of Powell, see Dupree (1957).

Waterways Commission (1908) recommended creation of a coordinated federal effort to plan for river development, in cooperation with state and local governments. However, in passing the Federal Water Power Act in 1920, Congress decided to limit the federal role in hydropower development to licensing nonfederal development of water power on navigable streams in a manner that would best promote comprehensive development of the water resources (Holmes, 1972). It gave this licensing authority to the newly created Federal Power Commission and directed the commission to conduct nationwide surveys of water-power-development opportunities.

Simultaneously, Congress was expanding the role of the Corps. It continued to fund navigation improvements through periodic Rivers and Harbors acts, and it passed other legislation increasing the Corps' flood-control responsibilities. Prior to 1936, the Corps' authority with respect to flood control had been limited to surveys and planning except on the lower Mississippi River. With the Flood Control Act of 1936, Congress committed to a national program for the control of floods and granted the Corps authority to survey, plan, and construct flood-control works throughout the nation.

With the coming of the Great Depression and the advent of the New Deal, the federal role was expanded in virtually all arenas. Multipurpose water development projects expanded rapidly as large-scale public-works programs became a favored means of providing employment and stimulating economic recovery. In addition to the substantial expansion of the construction activities of the U.S. Bureau of Reclamation (USBR) and the Corps, Congress created the Tennessee Valley Authority and added a soil and water conservation function to the U.S. Department of Agriculture (USDA). Large multipurpose regional water development projects occurred on the Missouri, Columbia, and Colorado rivers in addition to multibasin projects such as California's Central Valley Project. By the late 1930s, water projects accounted for 40 percent of the President's budget recommendations for public works (Holmes, 1972).

As the nation's attention turned to war in the early 1940s, economic recovery was at hand and the need for many New Deal programs declined. The National Resources Planning Board, which had been the lone mechanism for coordinating federal water programs during the 1930s, was abolished in 1941, leaving agencies with such programs free to compete for congressional funding. The political attractiveness of using federal funding to pay for expensive water development projects motivated Congress to authorize even more of these projects in the 1940s and 1950s than it had in the 1930s (Holmes, 1972). Most of the water project funding went to the Corps and the USBR. However, Congress started a new program within USDA during this period directed at smaller projects in rural, agricultural areas. Known as the Watershed Protection and Flood Prevention Act of 1954 (P.L. 83-566), this program authorized federal support for projects intended to reduce erosion, control flooding, and provide water supplies at a small watershed level.

Inevitably, various reactions to these projects set in. Budget concerns prompted demands for more thorough analysis of the economic benefits of the projects in relation to their costs. Critics noted the substantial subsidies these projects frequently provided to the direct beneficiaries and users. In the early 1950s environmentalists mounted a successful campaign to oppose construction of a dam at Echo Park in Dinosaur National Monument. Increasingly, the Truman and Eisenhower administrations and their Bureau of the Budget sought to force clearer economic justification for new projects. Circular A-47, issued at the end of the Truman administration, implemented a standard that proposed water projects would be expected to produce total benefits exceeding their costs. This standard had first appeared in the language of the Flood Control Act of 1936. Nevertheless, the funding of water projects had become a significant instrument of distributive politics as project beneficiaries, federal construction agencies, and members of Congress united in an "Iron Triangle" to secure a growing federal program of water projects (Ingram and McCain, 1977; Ingram, 1990).

The resulting search to define an appropriate federal role in the development and management of water resources resulted in the creation of a series of commissions and committees between 1946 and 1956 to make recommendations concerning a national water resources policy.[2] Then, in 1959 the Senate Select Committee on Water Resources was established and was chaired by Senator Robert Kerr of Oklahoma. This committee recommended federal action in five areas: streamflow regulation, water quality improvement, underground water storage, increased efficiency of water use, and increased water yield through desalting and weather modification. The primary rationale set forth for federally funded water projects was to enhance the supply of usable water, not necessarily to support local economic development. The Senate Select Committee also acknowledged a 1948 congressional finding that there was an appropriate federal role in the abatement of water pollution. There was also recognition that significant new federal funding would be required to address the nation's water quality needs. Finally, the Senate Select Committee report was also noteworthy because it addressed the need for an enhanced federal program of water resources research (U.S. Senate Committee on Interior and Insular Affairs, 1969).

The Era of Protection

In the decade of the 1970s fundamental changes occurred in national water policies and programs. A new federal agency, many new federal statutes, and new roles and involvement for various stakeholders and the public changed the

[2]Schad (1962) identifies eight such groups. Holmes (1972), pp. 40–43, provides a short summary of five of these: the first (1949) Hoover Commission; the 1950 President's Water Resources Policy Commission; the 1952 House Subcommittee to Study Civil Works; the second (1955) Hoover Commission; and the 1955 Presidential Advisory Committee on Water Resources Policy.

playing field for water resources and created new relationships. April 22, 1970, the first Earth Day, signaled a new level of interest in and concern about environmental issues, particularly those associated with water resources. Simultaneously, public involvement in all types of decision making related to resources and the environment greatly increased. Many new, active interest groups demanded a say in water resource policy issues. These groups also supported the conduct of research upon which new programs of environmental protection and enhancement could be based. In 1970 the U.S. Environmental Protection Agency (EPA) was created by Executive Order by President Richard Nixon, and it quickly became the major agency in the regulation and enhancement of water quality (following congressional actions described below). During this time Congress moved aggressively to place the federal government in a more central role as promoter and regulator of environmental protection.

The federal Water Pollution Control Act, originally enacted in 1948, was totally reshaped by the amendments of 1972 and 1977 (P.L. 92-500, P.L. 95-217) (Copeland, 2002). It ultimately became known as the Clean Water Act (CWA). This legislation declared all discharges into the nation's waters to be unlawful, unless such discharges were specifically authorized by permit. The act set ambitious objectives to restore and maintain the physical, biological, and chemical integrity of the nation's waters and to implement treatment of municipal and industrial wastewater ("point sources"), so that mandated standards could be met. Congress charged the EPA with determining the best available pollution control technologies for all major sources of discharges. Later amendments (1987) mandated the use of best management practices to control nonpoint sources of pollution.

In 1974, Congress enacted the Safe Drinking Water Act (SDWA, P.L. 93-523), which established national standards and treatment requirements for public water supplies, controls on underground injection of waste, protections for drinking water sources, and provisions for financing needed infrastructure. Congress enacted major amendments in 1986 to accelerate the schedule for regulating additional contaminants in water and to increase the protection of groundwater. Additional amendments in 1996 subsequently changed the way new contaminants would be addressed, focusing on new risk-based approaches, emphasizing the use of the best available science, and increasing the focus on pollution prevention through source water protection.

Other environmental legislation was also enacted during this period. The Resource Conservation and Recovery Act (RCRA) was passed in 1976, and the Comprehensive Environmental Response, Compensation, and Liability Act (CERCLA or "Superfund") was passed in 1980. Both of these laws had implications for new research on water quality and health issues. The Coastal Zone Management Act of 1972 linked the water quality protection efforts of EPA and the National Oceanic and Atmospheric Administration and required that coastal zone planning and management be coordinated with the CWA.

Through all these laws, EPA was directed, both explicitly and implicitly, to establish new national programs of research (1) on the effects of pollutants on human and ecological health and (2) on improved technologies and management approaches for the prevention and reduction of pollutants. The most recent directive came in 2002, when Congress added requirements for EPA to conduct research on the security of public water supplies and on prevention and response to terrorist or other attacks (Public Health Security and Bioterrorism Preparedness and Response Act of 2002, P.L. 107-188).

The national policy focus on protecting and enhancing the environment led to legislative mandates regarding activities that contributed to environmental degradation and the protection of unique ecosystems and species diversity, not all of which were based on regulation. Beginning in the mid 1980s, Congress included conservation provisions in the Farm Bill that were intended to minimize agriculture's role in degrading water quality and adversely impacting other environmental features. A Conservation Reserve Program was created that authorized payment to landowners who temporarily retired lands that were highly erodible or environmentally sensitive. The Wetlands Reserve Program made payments available to farmers who were willing to return croplands to wetlands for at least 30 years. The 1990 Farm Bill added provisions for retiring croplands adjacent to waterbodies that would be managed as filter or buffer strips. A Water Quality Incentive Program (now part of the Environmental Quality Incentives Program) offered technical and financial assistance to farmers willing to modify their agricultural practices in a manner that would reduce nonpoint source pollution.

In addition to this greatly heightened interest in water quality protection, other laws reflected a growing interest in protecting the scenic, recreational, and ecological values of water. In 1968 Congress passed the Wild and Scenic Rivers Act (P.L. 90-542), establishing a national system of rivers that would remain free of federal water development. The landmark Endangered Species Act of 1973 (P.L. 93-205) was intended to halt and reverse the trend toward increasing, human-caused extinction of plant and animal species. Because of the essential role played by aquatic and riparian environments in the life cycle of many species, water-related impairment of such environments has become a major focus of Endangered Species Act implementation. In 1986, Congress amended the Federal Power Act to require that issuance (or renewal) of licenses for hydroelectric power facilities give "equal consideration" to energy conservation; protection, mitigation of damage to, and enhancement of fish and wildlife (including spawning grounds and habitat); protection of recreational opportunities; and preservation of other aspects of environmental quality along with the traditional considerations of power and development (16 USC §797(f)).

FEDERAL SUPPORT OF WATER RESOURCES RESEARCH

The previous section illustrates that since the early 19th century, the nation's water policies have evolved from emphasizing navigation and settlement, to emphasizing physical development of water supplies to ameliorate scarcity, and finally to emphasizing the protection and enhancement of the environment. It is useful to consider how these policies were manifested in federally funded water resources research over the last 100 years.

The Beginnings

Some of the earliest examples of water resources research in the United States include the Gallatin Report of 1807, which detailed infrastructure conditions and needs of the nation's inland navigation routes, and the 1850s Ellet Report, which described conditions on the Mississippi River. Another early effort was the Humpheys and Abbot Report of 1861, which was recognized internationally for its pioneering research on the hydraulics of the Mississippi River. The next significant program of government-funded research specifically related to water was Powell's survey in the 1880s. The USGS gradually expanded its program of hydrographic surveys while individual researchers conducted investigations of groundwater, sediment transport, and water pollution (Langbein, 1981). Shortly after the turn of the century, the USBR began to measure flows in rivers in which it intended to construct storage facilities. The 1908 Inland Waterways Commission promoted the concept of comprehensive development of the nation's rivers, and the 1909 National Conservation Commission recommended large-scale hydrologic research in support of such comprehensive development (Holmes, 1972, p. 6). In the early decades of the 20th century, the role of the Corps expanded to include more planning and analysis related to its projects. Furthermore, the Federal Power Commission carried out surveys to determine potential locations for federally constructed hydroelectric power facilities, although authorization to construct and operate such facilities did not come until much later. The underlying purpose of all this federal research was to support water development and management programs and ensure that they contributed to regional economic development.

Concern about fisheries indirectly led to early research efforts on water topics. At the urging of Spencer Baird, a respected scientist who eventually directed the Smithsonian Institution, Congress established a Fish Commission in 1871 and made Baird the unpaid head (Dupree, 1957, p. 236). He established the Marine Biological Laboratory at Woods Hole, Massachusetts, with nonfederal funding in 1888. In 1903 the independent commission became the Bureau of Fisheries in the Department of Commerce and Labor. In somewhat similar fashion, Congress supported formation of a new division within USDA that eventually became the Division of the Biological Survey in 1896, with a program of research surveying the nation's biota (Dupree, 1957, pp. 238–239).

Water quality problems, especially related to drinking water, also motivated research efforts during the 19th century. Materials such as iron and new technologies such as steam-powered pumps, high-pressure systems, and filtration systems (slow sand, sand, and gravel) were first used to clean and distribute water (Rosen and Walker, 1968; Rosen, 1993; Webster, 1993; Embrey et al., 2002). By the middle of the century, several investigators in England and the United States had linked water contamination with certain infectious diseases. This finding spurred many remedial actions during the Sanitary Reform Movement, which was often effective in protecting public health even though it was not necessarily founded on scientific rationale (Rosen and Walker, 1968).

It was the discovery of bacteria and the development of scientific methods (not all of which were federally funded) that provided the objective bases for more advanced, technically based water treatment and distribution systems. In the 1860s and 1870s, the research of Louis Pasteur, Ferdinand Cohn, and Robert Koch led to significant conceptual breakthroughs and standardized scientific criteria and methods. These innovations laid the foundations of microbiology, which yielded new knowledge about bacteria and their roles in causing disease. Extensive public health benefits resulted (Rosen, 1993). In 1880, the German scientist Karl Eberth discovered the typhoid bacillus, leading to the linkage between polluted drinking water and typhoid fever (Goddard, 1966). By the late 1800s, many pathogens had been identified, public health laboratories such as the Lawrence Experiment Station in Massachusetts had been established, and experiments had been conducted to determine how bacteria could be killed. As knowledge grew about pathogens in drinking water, chlorination became the new standard for assuring safe water supplies for human consumption, spearheaded by American engineer Abel Wolman (Wolman and Gorman, 1931).

Several of the early federal research efforts occurred in the U.S. Public Health Service, which in 1910 conducted a two-year study of sewage pollution in streams around the Great Lakes and later investigated the role of pollution in the transmission of infectious disease. In 1913 the service established the Ohio River Investigation Station to conduct basic research on stream pollution and water purification. A pollution study of the Ohio River Basin, conducted jointly by the service and the Corps, served as the model for additional work authorized by the Water Pollution Control Act of 1948 (Goddard, 1966).

Funding for research on water problems came more slowly to other federal departments. Thus, for example, it was not until the 1930s that the USDA initiated a research program (through the Soil Conservation Service) to investigate techniques for controlling soil erosion. This early research evolved into the comprehensive water research program that is carried out by the USDA today, which focuses on the connection between agriculture and water resources.

The importance of federally sponsored scientific research to the country's World War II efforts prompted President Franklin Roosevelt in 1944 to ask Vannevar Bush to examine possible federal roles in supporting scientific research and development in the aftermath of the war. Bush's report (1950) concluded that the country would benefit greatly if the kind of government-sponsored research that had been so critical to the war effort was now brought to bear on important problems of public health and welfare. He recommended that federal support be provided for basic as well as applied scientific research. Bush's report is generally credited with providing the impetus for significant expansion of federally supported research in all areas of endeavor in the post-World War II period.

Post-World War II

In the immediate postwar period, water resources research continued at relatively modest levels and tended to be piecemeal among the federal agencies. It was focused primarily on issues related to the evaluation of water project proposals and improvements in planning techniques. The problems of water pollution and issues related to the management of fish and wildlife were also prominent. The 1948 Water Pollution Control Act increased the federal role in research related to water pollution and set the stage for even bigger increases that were to come 25 years later. In 1957 Congress appropriated funds for the USGS to establish for the first time a national-level program of core research related to hydrology.

The modern era of water resources research had its beginnings with the report of the Senate Select Committee on Water Resources (1961). This committee was the first to examine water resources research priorities in a comprehensive fashion. The committee recommended a coordinated scientific research program on water that would explore ways to increase available supplies and identify methods of increasing the efficiency of water use in the production of food and fiber and manufactured goods. The committee also noted the importance of expanding basic research into "natural phenomena" associated with water in all its forms. This latter finding would represent a significant broadening of scope for federally supported water research. Perhaps most important, the committee requested the executive branch to review existing water research programs and to develop a coordinated program of research aimed at meeting the needs identified in its 1961 report. The committee's work ultimately led to a broad, comprehensive vision of water resources research and was the first attempt to coordinate water resources research across the federal enterprise.

The newly elected John F. Kennedy and his administration responded by initiating studies at the National Academy of Sciences and the Federal Council for Science and Technology (FCST). Professor Abel Wolman prepared a report on water resources for the National Academy (NRC, 1962). In the report, he emphasized the need for more basic research related to water, stating that "less than one-fourth of one per cent of the total funds spent on water-resources development is allocated for basic research in water," and he identified a number of areas in need of additional research. The FCST Report to the President on Water Research (summarized in U.S. Senate Committee on Interior and Insular Affairs,

1969) acknowledged the need for increased research, both within and outside the federal agencies. It provided an inventory of existing agency research programs, and it proposed that federal research be coordinated by a committee of representatives from each relevant agency, chaired by a senior official. Thus, in late 1963, FCST established a Committee on Water Resources Research (COWRR), chaired by a representative of the President's Office of Science and Technology and including representatives from nine federal departments and commissions.[3] This committee was charged with:

- identifying technical needs and priorities in various research and data categories
- reviewing the overall programmatic adequacy in water resources research in relation to needs
- recommending programs and measures to meet these needs
- advising on desirable allocations of effort among the agencies
- reviewing/making recommendations on the manpower and facilities of the program
- recommending management policies and procedures to improve the quality and vigor of the research effort
- facilitating interagency communication and coordination at management levels

In 1966, COWWR published *A Ten-Year Program of Federal Water Resources Research* (COWWR, 1966, often referred to as The Brown Book). The committee defined two goals of the national water resources program. The first was "to manage our natural water resources and to augment them when necessary so as to meet all necessary requirements for water, both in quantity and quality." The second was "to minimize water-caused damages to life and property." The committee then specified the goal of federal water research to be the provision of knowledge necessary to meet the national water goals (listed above) as efficiently as possible. To accomplish these objectives, seven research areas were identified:

1. Develop methods for conserving and augmenting the quantity of water available.
2. Perfect techniques for controlling water to minimize erosion, flood damage, and other adverse effects.
3. Develop methods for managing and controlling pollution to protect and improve the quality of the water resource.

---

[3]The Office of Science and Technology and the Federal Council for Science and Technology transitioned in the 1970s to the Office of Science and Technology Policy in the Office of President and its Federal Coordinating Council for Science, Engineering, and Technology (FCCSET).

4. Develop and improve procedures for evaluating water resources development and management plans to maximize net socioeconomic benefits.
5. Understand the nature of water, the processes which determine its distribution in nature, its interactions with its environment, and the effects of man's activities on the natural processes.
6. Develop techniques for efficient, minimum-cost design, construction, and operation of engineering works required to implement the water resources development program.
7. Develop new methods for efficient collection of the field data necessary for the planning and design of water resources projects.

It is noteworthy that only one of these objectives (objective 5) mentions better understanding the environmental uses of water. This is symptomatic of the fact that The Brown Book was a creature of the era of development when augmentation of supply was the principal national strategy for addressing water scarcity. Environmental concerns had not yet manifested themselves on a national scale.

Within The Brown Book, existing federal water resources research was organized under eight topical headings, which were further subdivided into 44 areas (the original FCST subcategories, discussed in more detail in Chapter 4). In addition, 14 "major problem areas" needing additional research were identified. These included improved water planning, the ecological impacts of water development, problems of water pollution control, and the economics of water development. The report recommended nearly tripling the funding for water resources research by 1971.

The Senate Select Committee also had recommended comprehensive planning for water development in the nation's river basins, with the federal government taking the lead in cooperation with the states. Again, the Kennedy administration responded with strong interest, and the Water Resources Planning Act of 1961 was introduced in the 87th Congress. This bill proposed the creation of a Water Resources Council, comprised of the heads of the primary water-interested federal agencies. It proposed creating river basin commissions to carry out the planning. It offered the states substantial funding to carry out their own comprehensive water planning. In 1965 such a bill finally became law (P.L. 89-80).

The work of the Senate Select Committee had also elevated interest in promoting water research, leading to the enactment of the Water Resources Research Act of 1964. Inspired by the model of federally supported agricultural research stations in all 50 states, Title I of the act called for the creation of a Water Resources Research Institute in each state and Puerto Rico (eventually expanding to include the U.S. Virgin Islands, Guam, the District of Columbia, the Federated States of Micronesia, American Samoa, and the Commonwealth of the Northern Mariana Islands). It offered basic support of up to $100,000 per institute and authorized an institute-only competitive grants program requiring a 1:1 match. Title II created a competitive grants program, to be administered by the Secretary of the Interior, to encourage water research. An Office of Water Resources Research, reporting directly to the Secretary of the Interior, was established to administer the new law. In 1971 Congress increased the basic institute appropriation to $250,000. In 1974 the Office of Water Resources Research merged with the Office of Saline Water to form the Office of Water Research and Technology.

Ten years after the establishment of the Water Resources Research Institutes, the Congressional Research Service made an assessment of their effectiveness (Viessman and Caudill, 1976), which concluded that

> Considering the limited funding provided, the objectives of the Water Resources Research Act of 1964, as amended, have been met surprisingly well. Effective State water resources research centers have been established and have played an increasingly important role in State, regional, and national programs for water resources planning and development. A strong national research network is available, which has the potential for problem identification and solution prior to "crisis" situations. Funding levels have been meager, however, and unless increases are forthcoming, it is doubtful if the momentum achieved by the program can be sustained.

In addition to recommending increased funding, the report suggested establishing a national research strategy to guide research activities funded by the program, and it urged improved efforts to disseminate research results. Thus, in an effort to further strengthen the program, Congress passed the Water Research and Development Act of 1978 (P.L. 95-467), replacing the 1964 act. The new law required the creation of five-year research and development goals and objectives for the institutes. It required that funds provided to the institutes be at least 50 percent matched from nonfederal sources. It authorized an allotment of $150,000 per institute in 1979, increasing to $175,000 in 1980. In 1980, Congress authorized $150,000 per institute for 1981 and $160,000 for 1982.

Thus, COWWR, related agency research activities, the Water Resources Research Institutes, and the associated competitive grants program traced their origins to the Senate Select Committee. Most of the resulting research was focused on supporting physical water development activities.

The above-mentioned entities that sprang from the original Senate Select Committee were not the only ones to promote and manage water resources research. In the late 1960s, concerns arose over a number of proposals that entailed significant transbasin diversions. In response, Congress authorized the establishment of the National Water Commission to study the entire range of water resource issues and make recommendations on the scope and substance of a national water policy. The commission's report, which was issued in 1973, included discussion and recommendations about water resources research. Finding that water research had generally been successful at meeting past needs, the commission identified two concerns: (1) the need to develop closer ties between planning and research and (2) the need for a more "broadly based and intensive research and development effort to increase usable water supplies and to handle growing volumes of wastes" (National Water Commission, 1973). The commission organized federal water research into four general categories: (1) agency mission research, (2) earth science research, (3) research grant programs, and (4) research directed at new technologies. It found that about a third of the federal water research funding was going to water quality problems, followed by research related to the hydrologic cycle, water supply augmentation and conservation, and planning. It identified six areas of needed research, with priority being given to assessing impacts of water resources development, improving wastewater treatment, and evaluating water for energy production. (The other three areas were nonpoint source pollution, more efficient water use, and development of new technologies.) The report of the commission was completed coincident with events that led to the dissolution of the Nixon administration. For this reason, it was never formally transmitted to Congress and its recommendations languished.

As discussed earlier, the 1970s national strategy for managing water resources evolved away from a focus on physical development of water supplies toward environmental issues related to the development and management of water resources. Consistent with this evolution, the research agenda increasingly emphasized water quality and other environmental research to support the development and implementation of new environmental regulations. Congress authorized a number of significant programs of new research. Thus, for example, one section of the CWA directs EPA to conduct research on harmful effects of pollutants, effects of pesticides in water, effects of pollution on estuaries, the structure and function of freshwater systems, the effects of thermal discharges, and pathogen indicators in coastal recreational waters (33 USC § 1254). Moreover, EPA was to investigate ways to improve sewage treatment, improve water quality of lakes, and address oil pollution, waste oil, and agricultural pollution. Establishment of field laboratories was authorized in this provision, as was the creation of a Great Lakes water quality research program. Similarly, one section of the Safe Drinking Water Act authorized a general program of research related to drinking water protection, but it also specifically directed that research be undertaken on polychlorinated biphenyl contamination of drinking water, virus contamination of drinking water sources, and the reaction of chlorine and humic acids (42 USC § 300j-1). As part of both RCRA (42 USC § 6981-82) and CERCLA (42 USC § 9660), Congress established programs of research aimed at problems of solid and hazardous waste remediation, focusing on groundwater contamination.

These congressional directives focused research on the support of regulatory activities to maintain and enhance environmental quality. One result was that the broader and more coordinated and comprehensive multiagency water resources research agenda envisioned by COWRR and by the National Water Commission was ignored and ultimately abandoned. In the late 1970s and early 1980s, there were also important changes in the relationships between the states and the federal agencies. Both the CWA and SDWA established a federal–state relationship in which the federal agency set national standards related to the congressional agenda and the states were delegated the responsibility for implementing and enforcing the statutes. This approach fundamentally changed the role of the states from one primarily concerned with water management and development to one much more heavily focused on regulation, and it required that they devote additional resources to these activities.

COWRR, which had collected annual budget information on federal water resources research since 1965 and had survived several administrative reorganizations, included an unusually candid analysis of The Brown Book's effectiveness in its 1977 report (COWRR, 1977). It stated: "judging from the history of financial support devoted to water research as reported in COWRR annual reports, the long-range plan had little impact, even indirect, on the course of water research activities during the decade from 1966 to 1976." In particular, actual funding for water research had fallen well short of that recommended. Within the general categories, funding for research on the nature of water, for manpower grants and facilities, and for scientific and technical information declined significantly. Only the areas of water quality and resources data showed an increase in funding (likely caused by the creation of EPA and the passage of the CWA). The committee stated that "overall the only conclusion that can be reached is that the federal water research program has fared poorly in the Congress and with the Office of Management and Budget."

Concluding that the primary reason for this lack of success was a failure (1) to provide "an objective, defensible analysis of the critical national problems which involve water resources" and (2) to identify the "deficiencies in knowledge and understanding which must be eliminated for the development of effective resolution of these problems," COWRR offered a "refocused" water research program for the next five years (COWRR, 1977). Toward this end, six national issues motivating the need for water research were identified: energy, food and fiber production, the environment and public health, population growth, land use, and materials. The water problems inherent in these issues suggested six general research areas: hydrologic and hydraulic processes, water quality, planning and institutions, atmospheric and precipitation processes, hydrologic–ecological relationships, and water supply development and management. Under each of these topics, COWRR identified areas that should receive much greater attention over the next five years. In addition, COWRR highlighted the importance of financial support for data collection and for manpower and training. Finally, the report made six recommendations: (1) substantially increase funding, (2) increase efforts to develop a unified national program of water resources research, (3) better balance mission agency research programs between in-house and outside research, (4) study manpower needs for the water resources field, (5) give careful attention to COWRR's suggested areas of research in formulating federal agency research programs, and (6) coordinate the programs of the Intergovernmental Committee on Atmospheric Sciences with the water resources research community.

COWRR was abolished during a far-reaching reorganization of its parent body FCCSET in 1977. There was nonetheless an interest in sustaining the federal role in water resources research, as evidenced by the 1978 Water Research and Development Act, which called for the Secretary of the Interior to prepare a five-year water resources research plan.[4] The Department of Interior enlisted the support of the National Research Council (NRC) to review a draft of the five-year plan in 1980. In now familiar fashion, the NRC committee provided yet another view of the nation's water problems and the related needs for research, and it offered its own classification system for water research, according to the five categories below (NRC, 1981):

- Category I: atmospheric, hydrologic, and hydraulic processes
- Category II: ecological and environmental relationships in water resources
- Category III: water-quality protection and control
- Category IV: water resources management
- Category V: institutional analysis

The NRC committee then presented its research priorities under each of the five categories.

Because the COWRR practice of collecting annual budget data on water resources research ended in 1975, the NRC committee had limited information from which to draw conclusions about the adequacy of funding in each of these five categories. Thus, the NRC committee was compelled to rely mainly on ad hoc explanations of each agency's research. Nevertheless, it was able to conclude that funding was inadequate for Categories I, II, and V; excessive for Category III; and adequate for Category IV. Among its many other suggestions, the NRC committee noted the need to address research priorities and to discuss the linkage between water problems and research, the need to provide for interagency coordination and elimination of duplication (for which it found no evidence), and the need to address the policy issues implicit in many aspects of water research. Finally, the NRC committee offered three alternative organizational arrangements: a managed multiagency research program to be operated by an independent office, the creation of an interagency committee in the Office of Science and Technology Policy, or placing organizational responsibilities within the Water Resources Council.

[4]The act directed the secretary to seek cooperation and advice from federal agencies, state and local governments, and private institutions and individuals to ensure that suggested research will supplement and not duplicate existing research, will stimulate research and development in needed areas, and will establish a comprehensive, nationwide program of water resources research and development.

Water Resources Research from the 1980s and Beyond

The administration of President Ronald Reagan, particularly in its early years, espoused a limited role for the federal government in many spheres, including scientific research. Thus, for example, the 1981 Science and Technology Report to Congress (OSTP, 1981) stated:

> The inception of the new Administration's programs early in 1981 brought a philosophical change to the natural resources area. Rather than trying to solve national problems through extensive Federal programs, the decision was made to rely wherever possible on the private sector for natural resources development.
>
> A second guiding premise of the Administration's policies is that many functions previously held by Federal agencies really belong to State and local governments. For example, the States will be expected to support research efforts dealing with their own water resource problems and development projects.

More specifically, the Reagan administration defined its policy on water resources research rather narrowly. The policy had three goals: (1) enhancing the capability of state governments to manage water, (2) encouraging nonfederal investment in water-related research, and (3) building collective national technical capability to solve water problems. In addition, the administration acknowledged that there was a federal role in coordinating and facilitating the flow of information generated from research to state water resources managers, planners, and policy makers. Moreover, it supported the proposition that the federal government should continue to fund important basic research and continue to carry out programs mandated by statute while seeking to transfer these activities to the states where appropriate (Tom Bahr, Office of Water Policy between 1982 and 1984, personal communication, 2003).

In 1981 the Reagan administration's Office of Management and Budget decided not to request funding for either the state Water Resources Research Institute program or the competitive grants program under the Water Resources Research Act. Congress ultimately funded the institute program, but at a reduced level, and it elected not to fund the competitive grants program. In 1982, Secretary of the Interior James Watt abolished the Office of Water Research and Technology, placing the institute program under an Office of Water Policy in the Department of the Interior and the matching grants program under the USBR. In its fiscal year 1984 Appropriations Act, Congress recommended that the institute program be placed under the USGS, where it remains to this day. In 1984, the Reagan administration again opposed reauthorization of the Water Resources Research Institute and competitive grants programs, arguing that such research should be financed by the beneficiaries, whom they identified as the states and industry. Despite this opposition, Congress reauthorized the programs and even overrode a presidential veto of the bill (Water Resources Research Act of 1984, P.L. 98-242).

Concerned by the Reagan administration's move to reduce the federal role in water research, the Universities Council on Water Resources obtained a grant from the National Science Foundation to hold a National Conference on Water Resources Research. In total, 15 papers were prepared, circulated for review, and discussed at a conference in 1985. In the conference proceedings (Universities Council on Water Resources, 1985), three "immediate needs" were identified: (1) improved coordination among universities, federal and state agencies, and the private sector, (2) strengthened organizational and fiscal arrangements, and (3) a regular review of expenditures and priorities. According to the report, there was "substantial agreement" that a committee of the NRC should recommend a national water research agenda. Participants concluded that it was difficult to make an adequate assessment of existing research because of the absence of standardized categories of problems and associated research and because of the absence of uniform definitions. Considerable discussion centered on the relationship between policy and research, how to make research relevant to policy users, and how to ensure that research results are transmitted in usable forms. The report emphasized the need for education and training to be considered a fundamental part of a national water research agenda. Work groups proposed an extensive agenda of important water research priorities for the next three to five years, and the report included six "themes" that run through this agenda.[5]

The administration of President George Bush continued to view water resources research in the context of specific agency responsibilities, such as water quality protection, or in relation to larger problems of which water was a part, such as global change. In 1990, Congress passed and the President signed the Global Change Research Act of 1990 (P.L. 101-606), the purpose of which was

> to require the establishment of a United States Global Change Research Program aimed at understanding and responding to global change, including the cumulative effects of human activities and natural processes on the environment, to promote discussions towards international protocols in global change research, and for other purposes.

The Bush administration continued the Reagan policy of opposing federal funding for state Water Resources Research Institutes. However, Congress continued to affirm its support, and in 1990 Congress authorized funding of $10 million annually. Congress also reinstituted a regional competitive grants program requiring a 1:1 match.

There have been other significant events in the world of water resources research, although they have been more sporadic and on a much smaller scale

---

[5]These are (1) the need for critical information, (2) the importance of institutional research, (3) the importance of financing, budgeting, and pricing, (4) scientific understanding of water regimes, (5) technology improvements, and (6) ecosystem research.

than what occurred during the 1960s and 1970s. The NRC's Water Science and Technology Board produced a 1991 report *Opportunities in the Hydrologic Sciences* "to help guide science and educational policy decisions and to provide a scientific framework and research agenda for scientists, educators, and students making career plans." The report made the case for viewing hydrologic science as a multidisciplinary field that focuses on water's role in many of the physical, chemical, and biological processes regulating the earth's system, noting that the "field needed sounder scientific underpinnings, particularly as we begin to take a more global and system-oriented view of our environment" (NRC, 1991). In response, the National Science Foundation created a new research program for hydrologic sciences within the Geosciences Directorate (James, 1995). This effort was one of the first to consider water resources research as a basic rather than applied science.

Similarly, in 1994, EPA and the National Science Foundation initiated a Water and Watersheds research grant program, the purpose of which was to "synthesize physico-chemical, biological, and social science expertise in addressing water and watershed issues." Impetus for this program originated with *The Freshwater Imperative: A Research Agenda* (Naiman et al., 1995). Asserting that freshwater ecosystems are the "central component" of regional and global sustainability, this report identified four primary areas for research: restoring and rehabilitating ecosystems, maintaining biodiversity, understanding the effects of modified hydrologic regimes, and describing the importance of ecosystem goods and services provided by freshwater ecosystems. The report highlighted the importance of managing freshwater systems on the basis of "integrative and accurate measures of human and environmental conditions." USDA joined the Water and Watersheds program as a supporter in 1996. The number of research proposals routinely overwhelmed the level of funding support during the six-year life of this program.

Efforts to support water resources research within the President's office have also been variously resurrected over the years. In 1993 President Bill Clinton established the National Science and Technology Council (NSTC) to serve as a cabinet-level mechanism for coordinating science, space, and technology policies across the federal government. The Committee on Environment and Natural Resources (CENR), one of nine committees under the NSTC, was charged with improving coordination among federal agencies involved in environmental and natural resources research and development, establishing a strong link between science and policy, and developing a federal environmental and natural resources research and development strategy that responds to national and international issues. Using interagency task teams, CENR fostered research reports on two water-related topics: hypoxia in the Gulf of Mexico and science needs for Pacific salmon restoration (NSTC, 2000).

In 1995 the Clinton administration elected not to request funding for the Water Resources Research Institute program. The administration's formal position was that the federal government should not fund programs that are not inherently federal responsibilities. However, it appeared that no administration had an incentive to support the program, given that Congress routinely added the necessary funding with or without administration support. The regional competitive grants program was also continued at modest levels.

In the last decade, it has been recognized that resolving most of the major contemporary water problems goes beyond the capability of any single federal or nonfederal organization. Thus, multiagency, comprehensive approaches to both place-based and generic topics in water resources research have been supported to address priority problems. Examples include studies of the Chesapeake Bay and the Great Lakes, the National Acid Precipitation Assessment Program, and the general problem of nonpoint source pollution. The Water Quality 2000 program of the Water Environment Federation issued *A National Water Agenda for the 21st Century* (WEF, 1992), which lays out various strategic options developed by representatives of more than 80 public, private, and nonprofit organizations. The National Nonpoint Source Forum was another public, private, and nonprofit initiative convened by the National Geographic Society and The Conservation Fund to develop partnership approaches to mitigate nonpoint source problems (National Geographic Society and The Conservation Fund, 1995).

In 1998, partly in response to the 25th anniversary of the CWA, EPA and the USDA jointly issued a Clean Water Action Plan at the direction of President Clinton and Vice President Al Gore (EPA, 1998). This plan involved input from, and identified key actions for, *all* federal agencies whose mission relates to water, including Agriculture, Commerce, Defense, Energy, Interior, Justice, Transportation, EPA, and the Tennessee Valley Authority. The plan calls for efforts to enhance watershed protection and strengthen ways to reduce polluted runoff, including a focus on new research. Although the plan garnered wide public support, congressional action was not forthcoming.

More recently, the Water Science and Technology Board of the NRC produced a report entitled *Envisioning the Agenda for Water Resources Research in the Twenty-first Century* (NRC, 2001). *Envisioning,* which forms the basis for the current report, outlined 43 research priorities and called for the creation of a "national water research board" to establish and oversee the national water research agenda. To guide such a board, the report offered the following principles:

- An effective alliance with and active participation of water resources research stakeholders is required.
- A systematic, strategic, and balanced agenda of both core and problem-driven research priorities should be set to meet short- and long-term needs. The core research agenda should develop (1) greater understanding of the basic processes—physical, biological, and social—that underlie environmental systems at different scales, (2) appropriate environmental monitoring programs, and (3) research tools to identify and measure structural and functional attributes of aquatic and related ecosystems.

- The national water resources research effort should be coordinated to reduce needless duplication and to ensure that gaps do not occur.
- The research effort should be multidisciplinary and interdisciplinary.
- The research effort should be proactive and anticipate the nation's needs and the environmental impacts of management options.
- The research effort should be accountable to the public to ensure that the water resources research investment has been appropriately utilized to meet the nation's needs.

Finally, perhaps the latest recognition of the need for coordination to tackle the nation's water problems is the 21st Century Water Commission Act of 2003. Introduced in the 108th Congress in January 2003 as H.R. 135, the proposal would establish the 21st Century Water Commission to study and develop recommendations for a comprehensive water strategy to address future water needs, particularly to ensure an adequate and dependable supply of water to meet U.S. needs for the next 50 years. As of this writing, this legislation has been adopted by the House of Representatives and awaits Senate action.

SUMMARY

Figure 2-1 summarizes the seminal events in water resources research that have been discussed in this chapter. Federal support of water-related research developed slowly because of the prevailing view during much of the 1800s that science was not a governmental function. As federal involvement in the development of rivers for navigation, flood control, and storage of water for irrigation grew, so did accompanying research (although in the early 20th century, the need was primarily for engineers, not scientists). Nevertheless, federal scientists played an important early role in the collection of information about the extent of the nation's water resources, the nature of groundwater, and the need for protecting drinking water for public health purposes.

It was not until the 1950s that Congress committed itself to supporting a comprehensive program of water research. The commitment, which was short-lived, peaked during the 1960s when Congress and the executive branch achieved a consensus in developing and funding a comprehensive research program and in coordinating its implementation. During this period, the two branches of government shared the view that the federal role in water entailed funding its development for human use while reducing problems of pollution. By the 1970s, the growing interest in environmental protection conflicted with interests in water development, such that the policy consensus was splintered. This cast the federal government into more of a regulatory role and deemphasized the federal role in promoting economic growth through water resources development.

As broad support for national water policies that focused on development began to erode, competing interests pursued their individual objectives. Begin-

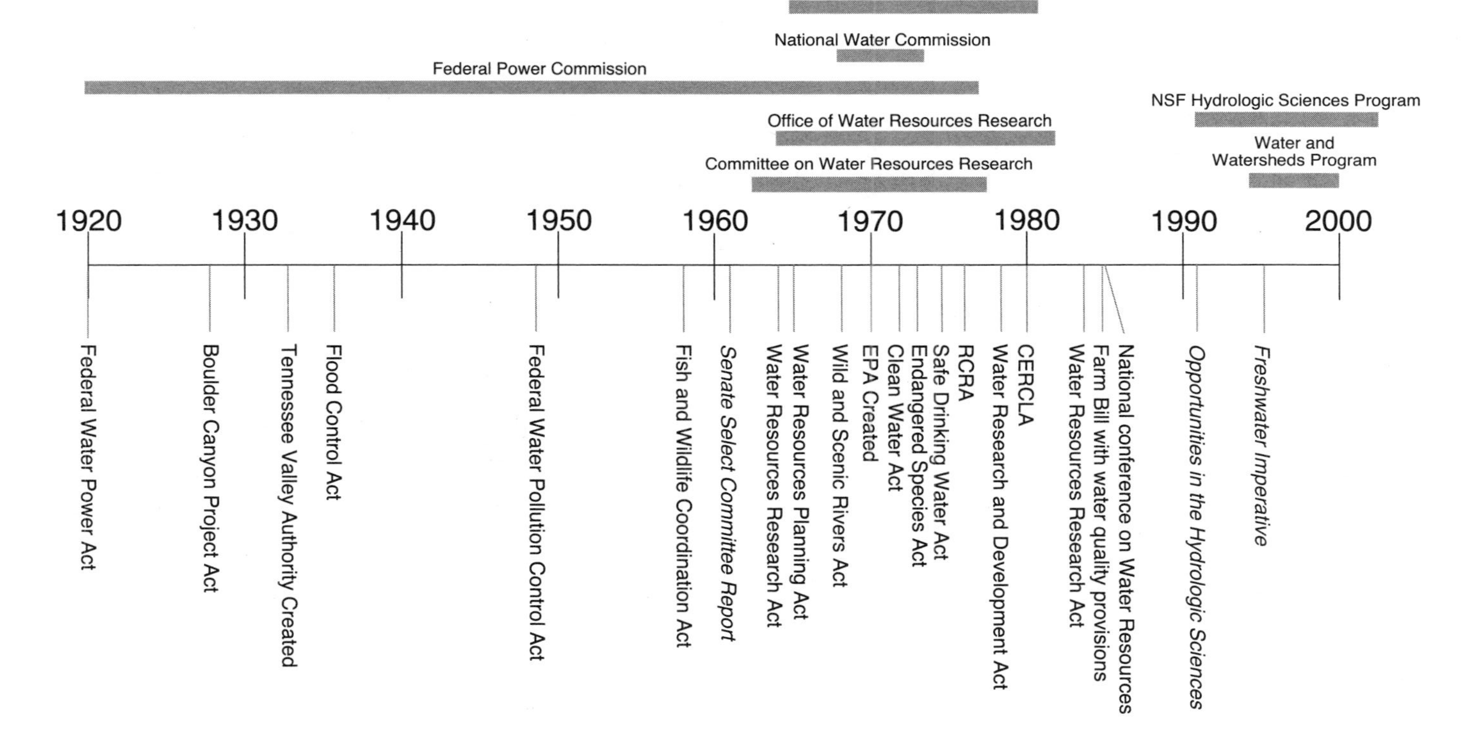

FIGURE 2-1 Timeline of 20th-century events, legislation, and publications in water resources research that are mentioned in this chapter.

ning in the 1970s, water research became tied to programmatic "thrusts" of administrations or to statutorily defined objectives of Congress. In the 1980s and 1990s, the Reagan, Bush, and Clinton administrations asserted a more limited federal role in water resources research. In their view, research should be closely connected to helping to meet federal agency missions or to addressing problems beyond the scope of states or the private sector (such as global climate change or hypoxia in the Gulf of Mexico). Congress, on the other hand, generally supported a broader approach to water resources research, but one that it could actively supervise through the legislative and appropriations processes. A consequence of the devolving of responsibility for water resources research back to the states was the neglect of long-term, basic research in favor of applied research that would lead to more immediate results.

As summarized in Box 2-1, the priority elements of a national water resources research agenda have been identified in widely varying ways by many of the organizations and reports identified in this chapter. In some respects, each agenda reflects the view of the federal government's role that was prevalent at the time the agenda was created. Thus, early research agendas stress research that would assist comprehensive water development, balanced with an interest in better decision-making criteria for determining whether such development warranted federal support. Later, as political support for federal funding of water development weakened and as the federal role shifted to technological and regulatory protection of water quality, the emphasis of the research agendas shifted accordingly. No doubt these variations also reflect to some degree the mental frameworks and particular interests of those who developed the agendas.

And yet the general topics of scientific concern found in the agendas of Box 2-1 remain remarkably similar: water-based physical processes; availability of water resources for human use and benefit, including improving and protecting water quality; and hydrology–ecology relationships. The reappearance of the same topics over and over suggests that the nation's research programs, both individually and collectively, have not responded in an adequate manner. Box 2-1 furthermore suggests that there is no structure in place to make use of the research agendas generated by various expert groups. Indeed, at the national level there is no coordinated process for considering water resources research needs, for prioritizing them for funding purposes, or for evaluating the effectiveness of research activities. It is no surprise that common refrains within many of the reports cited in Box 2-1 are for better coordination of research efforts—a topic that is returned to in Chapter 6.

There could be several explanations for why the country has failed to mount a serious, comprehensive water resources research program in spite of more than half a dozen efforts to define national research agendas in the past 40 years. The responsibilities for water resources development and management are fragmented among a number of agencies, and it appears that the agencies have no incentive to act in concert with each other to support the development of a unified national

BOX 2-1
Water Research Priorities of Various Organizations, Committees, and Reports

Senate Select Committee Report (1961)

- expansion of basic water research
- more balanced and better-constructed program of applied research for increasing water supplies
- an expanded program of applied research for conservation and making better use of existing supplies
- evaluation of completed projects with a view to making them more effective in meeting changing needs and providing better guidelines for future projects
- federally coordinated research programs to meet these objectives

Abel Wolman Report (NRC, 1962)

Arid areas research:

- conjunctive ground-surface water management
- evaporation suppression and transpiration control
- salinity control and use of saline water
- factors governing the entrainment, transport, and deposition of suspended sediment
- factors governing snow melt
- induced rainfall

Humid areas research:

- developing water-purification methods
- means of forecasting the effects of wastes on receiving water and toward quantifying pollution damages
- means for the detection and identification of traces of pollutants and toxicological research on their possible chronic effects on public health

All areas research:

- forecasting and controlling channel modifications
- improving the process of approximating optimum water resources systems
- improving streamflow forecasting
- improving weather forecasting
- physiological aspects of water

Committee on Water Resources Research (1966) (The Brown Book)

- research on water resources planning
- research on water pollution control
- research on water conservation

continued

BOX 2-1 Continued

- ecological impact of water development
- effect of man's activities on water
- costs of water resources development
- research on "far-out" ideas
- the problem of climatic change
- information storage and dissemination
- a program of problem assessment
- water resources research laboratories
- experimental watershed studies
- coordination of research
- manpower

National Water Commission (1973)

- assessing impacts of water resources development
- improving wastewater treatment
- evaluating water for energy production
- nonpoint source pollution
- more efficient water use
- development of new technologies

Committee on Water Resources Research (1977)

The committee identified six national issues motivating the need for water research: energy, food and fiber production, the environment and public health, population growth, land use, and materials. The water problems inherent in these issues suggested six general research areas:

- hydrologic and hydraulic processes
- water quality
- planning and institutions
- atmospheric and precipitation processes
- hydrologic–ecological relationships
- water supply development and management

NRC Federal Water Resources Research (NRC, 1981)

I. Physical processes

- hydrologic characteristics of vadose zone
- atmospheric transport and precipitation of contaminants
- flood frequency determination
- hydrologic factors in water quality
- climate variability and trends
- erosion, sedimentation, and nutrient transport
- weather and hydrologic forecasting

continued

II. Ecological–environmental

- effects of waterborne pollutants on aquatic ecosystems
- consequences of waste disposal in marshes, estuaries, and oceans
- physical alteration of wetlands and estuaries
- environmental degradation from water projects

III. Water quality

- significance of trace contaminants to human health
- water reuse
- control of contaminants from energy development
- land disposal of wastes
- monitoring for pollution control

IV. Water management

- water problems of food and fiber production in stressed environments
- conjunctive management of ground and surface water
- water conservation in municipal, industrial, energy, and agricultural uses
- control of pollution from nonpoint sources
- management systems for water resources
- management of resources under flood and drought hazards

V. Institutional

- institutional arrangements for reallocation of water
- institutional arrangements for groundwater management
- assignment of responsibilities for water and related resource management among federal, state, and local levels of government
- institutional arrangements for water conservation
- flood and drought hazard mitigation
- resolution of conflicts over alternative courses of action
- institutional arrangements for achieving erosion and sediment control
- impacts of water management policies and programs
- institutional arrangements for water resources research

Opportunities in the Hydrologic Sciences (NRC, 1991)

- chemical and biological components of the hydrologic cycle
- scaling of dynamic behavior
- land surface–atmospheric interactions
- coordination of global-scale observations of water reservoirs and the fluxes of water and energy
- hydrologic effects of human activity
- maintenance of continuous long-term datasets
- improved information management
- interpretation of remote sensing data
- dissemination of data from multidisciplinary experiments

continued

BOX 2-1 Continued

The Freshwater Imperative (Naiman et al., 1995)

- restoring and rehabilitating freshwater ecosystems
- maintaining biodiversity
- understanding the effects of modified hydrologic flow patterns
- describing the importance of ecosystem goods and services provided by freshwater ecosystems
- developing new paradigms of predictive management based on interdisciplinary research and new models of institutional organization that can respond to novel or unforeseen problems

Envisioning the Agenda for Water Resources Research (NRC, 2001)

Water availability

- develop new and innovative supply-enhancing technologies
- improve existing supply-enhancing technologies such as wastewater treatment, desalting, and groundwater banking
- increase safety of wastewater treated for reuse as drinking water
- develop innovative techniques for preventing pollution
- understand physical, chemical, and microbial contaminant fate and transport
- control nonpoint source pollution
- understand impact of land-use changes and best management practices on pollutant loading to waters
- understand impact of contaminants on ecosystem services, biotic indices, and higher organisms
- understand assimilation capacity of the environment and time course of recovery following contamination
- improve integrity of drinking water distribution systems
- improve scientific bases for risk assessment and risk management with regard to water quality
- understand national hydrologic measurement needs and develop a program that will provide these measurements
- develop new techniques for measuring water flows and water quality, including remote sensing and *in situ* techniques
- develop data collection and distribution in near real time for improved forecasting and water resources operations
- improve forecasting the hydrologic cycle over a range of time scales and on a regional basis
- understand and predict the frequency and cause of severe weather (floods and droughts)
- understand recent increases in damage from floods and droughts
- understand global change and its hydrologic impacts

continued

Water use

- understand determinants of water use in the agricultural, domestic, commercial, public, and industrial sectors
- understand relationship of agricultural water use to climate, crop type, and water application rates
- develop improved crops for more efficient water use and optimize the economic return for water used
- develop improved crop varieties for use in dryland agriculture
- understand water-related aspects of the sustainability of irrigated agriculture
- understand behavior of aquatic ecosystems in a broad, systematic context, including their water requirements
- enhance and restore species diversity in aquatic ecosystems
- improve manipulation of water-quality parameters to maintain and enhance aquatic habitats
- understand interrelationship between aquatic and terrestrial ecosystems to support watershed management

Water institutions

- develop legal regimes that promote groundwater management and conjunctive use of surface and groundwater
- understand issues related to the governance of water where it has common pool and public good attributes
- understand uncertainties attending to Native American water rights and other federal reserved rights
- improve equity in existing water management laws
- conduct comparative studies of water laws and institutions
- develop adaptive management
- develop new methods for estimating the value of nonmarketed attributes of water resources
- explore use of economic institutions to protect common pool and pure public good values related to water resources
- develop efficient markets and market-like arrangements for water
- understand role of prices, pricing structures, and the price elasticity of water demand
- understand role of the private sector in achieving efficient provision of water and wastewater services
- understand key factors that affect water-related risk communication and decision processes
- understand user-organized institutions for water distribution, such as cooperatives, special districts, and mutual companies
- develop different processes for obtaining stakeholder input in forming water policies and plans
- understand cultural and ethical factors associated with water use
- conduct *ex post* research to evaluate the strengths and weaknesses of past water policies and projects

water resources research agenda. Furthermore, over the last 40 years the competition for federal funds in general and research funding in particular has intensified, with water resources research not being a national priority compared to health and defense-related issues. In the face of historical failures to mount an effective, broadly conceived program of national water resources research, it is reasonable to ask "Why bother with yet another comprehensive proposal?" The answer lies in the sheer number of water resources problems (as illustrated in Chapter 1) and the fact that these problems are growing in both number and intensity. If the nation is to address these problems successfully, an investment must be made not only in applied research but also in fundamental research that will form the basis for applied research a decade hence. A repeat of failed past efforts will likely lead to enormously adverse and costly outcomes on the status and condition of water resources in almost every region of the United States.

REFERENCES

Bush, V. 1950. Science–The Endless Frontier (40th Anniversary Edition). Washington, DC: The National Science Foundation.

Committee on Water Resources Research (COWWR). 1966. A Ten-Year Program of Federal Water Resources Research. Washington, DC: Federal Council for Science and Technology, Office of Science and Technology, Executive Office of the President.

Committee on Water Resources Research (COWWR). 1977. Directions in U.S. Water Research: 1978–1982. Washington, DC: Federal Coordinating Council for Science, Engineering and Technology.

Copeland, C. 2002. Clean Water Act: A Summary of the Law. CRS Report for Congress, RL30030, updated January 24, 2002. Washington, DC: Congressional Research Service.

Dupree, A. H. 1957. Science in the Federal Government: A History of Policies and Activities to 1940. Cambridge, MA: Belknap Press of Harvard University Press.

Embrey, M., R. Parkin, and J. Balbus. 2002. Handbook of CCL Microbes in Drinking Water. Denver, CO: American Water Works Association.

Environmental Protection Agency (EPA). 1998. Clean Water Action Plan: Restoring and Protecting America's Waters. Washington, DC: EPA.

Goddard, M. K. 1966. Water Supply and Pollution Control. *In* Origins of American Conservation, Henry Clepper (ed.). New York, NY: Ronald Press.

Hill, F. G. 1957. Roads, Rails and Waterways: The Army Engineers and Early Transportation. Norman, OK: University of Oklahoma Press.

Holmes, B. H. 1972. A History of Federal Water Resources Programs, 1800–1960. Miscellaneous Publication No. 1233. Washington, DC: USDA Economic Research Service.

Ingram, H. 1990. Water Politics: Continuity and Change. Albuquerque, NM: University of New Mexico Press.

Ingram, H., and J. R. McCain. 1977. Federal Water Resources Management: The Administrative Setting. The Public Administration Review 37(5). September/October.

Inland Waterways Commission. 1908. Preliminary Report, Doc. 325, 60th Cong., 1st sess., 1908 at 27. Washington, DC: Inland Waterways Commission.

James, L. D. 1995. NSF research in hydrologic sciences. Journal of Hydrology 172:3–14.

Langbein, W. D. 1981. A History of Research in the USGS/WRD. WRD Bulletin, October–December.

Maass, A. 1951. Muddy Waters: The Army Engineers and the Nation's Rivers. Cambridge, MA: Harvard University Press.

Naiman, R. J., J. J. Magnuson, D. M. McKnight, and J. A. Stanford, eds. 1995. The Freshwater Imperative: A Research Agenda. Washington, DC: Island Press.

National Geographic Society and The Conservation Fund. 1995. Water: A Story of Hope. Washington, DC: The Terrene Institute.

National Research Council (NRC). 1962. Water Resources: A Report to the Committee on Natural Resources of the National Academy of Sciences—National Research Council, Publication 1000-B. Washington, DC: National Academy Press.

National Research Council (NRC). 1981. Federal Water Resources Research: A Review of the Proposed Five-Year Program Plan. Washington, DC: National Academy Press.

National Research Council (NRC). 1991. Opportunities in the Hydrologic Sciences. Washington, DC: National Academy Press.

National Research Council (NRC). 2001. Envisioning the Agenda for Water Resources Research in the Twenty-first Century. Washington, DC: National Academy Press.

National Science and Technology Council (NSTC). 2000. 2000 Annual Report, at p. 14. Washington, DC: Office of Science and Technology Policy NSTC.

National Water Commission. 1973. Water Policies for the Future. Washington, DC: National Water Commission.

Office of Science and Technology Policy (OSTP). 1981. In Cooperation with the National Science Foundation, Annual Science and Technology Report to the Congress. Washington, DC: OSTP.

Pisani, D. J., 1992. To Reclaim a Divided West: Water, Law, and Public Policy 1848–1902. Albuquerque, NM: University of New Mexico Press.

Rosen, G. 1993. The History of Public Health. Baltimore, MD: Johns Hopkins University Press.

Rosen, G., and M. E. M. Walker. 1968. Pioneers for Public Health. Freeport, NY: Books for Libraries Press.

Schad, T. M. 1962. An analysis of the work of the Senate Select Committee on National Water Resources, 1959–1961. Natural Resources Journal 2:226–247.

Senate Select Committee on National Water Resources. 1961. Senate Report No. 29, 87th Cong., 1st Sess.

U. S. Senate Committee on Interior and Insular Affairs. 1969. History of the Implementation of the Recommendations of the Senate Select Committee on National Water Resources. Washington, DC: U.S. Government Printing Office.

Universities Council on Water Resources. 1985. Summary Report of the National Conference on Water Resources Research. Lincoln, NE.

Viessman, W., Jr., and C. K. Caudill. 1976. The Water Resources Research Act of 1964: An Assessment. Washington, DC: Committee on Interior and Insular Affairs, 94th Congress.

Webster, C. (ed.). 1993. Caring for Health: History and Diversity. Norwich, UK: Open University.

Water Environment Federation (WEF). 1992. A National Water agenda for the 21st Century. Alexandria, VA: WEF.

Wolman, A., and A. E. Gorman. 1931. The Significance of Waterborne Typhoid Fever Outbreaks, 1920–1930. Baltimore, MD: Williams and Wilkins Co.

Worster, D. 2001. A River Running West: The Life of John Wesley Powell. New York: Oxford University Press.

3

Water Resources Research Priorities for the Future

The pressing nature of water resource problems was set forth in Chapter 1. The solution to these problems is necessarily sought in research—inquiry into the basic natural and societal processes that govern the components of a given problem, combined with inquiry into possible methods for solving these problems. In many fields, descriptions of research priorities structure the ways in which researchers match their expertise and experience to both societal needs and the availability of research funding. Statements of research priorities also evolve as knowledge is developed, questions are answered, and new societal issues and pressures emerge. Thus, the formulation of research priorities has a profound effect on the conduct of research and the likelihood of finding solutions to problems.

Statements of research priorities developed by a group of scientists or managers with a common perspective within their field of expertise can have a relatively narrow scope. Indeed, this phenomenon has resulted in numerous independent sets of research priorities for various aspects of water resources. This has come about because water plays an important role in a strikingly large number of disciplines, ranging from ecology to engineering and economics—disciplines that otherwise have little contact with each other. Thus, priority lists from ecologists emphasize ecosystem integrity, priority lists from water treatment professionals emphasize the quantity and quality of the water supply, and priority lists from hydrologists emphasize water budgets and hydrologic processes. In recent years, the limitations of discipline-based perspectives have become clear, as researchers and managers alike have recognized that water problems relevant to society necessarily integrate across the physical, chemical, biological, and social sciences. Narrowly conceived research produces inadequate solutions to such problems;

these in turn provide little useful guidance for management because critical parts of the system have been ignored. For example, the traditional subdivision of water resource issues into those of quality and quantity is now seen as inadequate to structure future research, given that water quality and quantity are intimately, causally, and mechanistically connected. Similarly, theoretical studies of water flows (hydrology) and aquatic ecosystems (limnology) can no longer be viewed as independent subjects, as each materially affects the other in myriad ways. Finally, the physical, chemical, and biological aspects of water cannot adequately be investigated without reference to the human imprint on all facets of the earth's surface. Thus, the challenge in identifying water resources research needs is to engage researchers in novel collaborations and novel ways of perceiving the research topics that they have traditionally investigated.

Water resources research priorities were recently extensively considered by the Water Science and Technology Board (WSTB) in *Envisioning the Agenda for Water Resources Research in the Twenty-first Century* (NRC, 2001a). This resulted in a detailed, comprehensive list of research needs, grouped into three categories (Table 3-1); the reader is referred to NRC (2001a) for a detailed description of each research need. The category of *water availability* emphasizes the interrelated nature of water quantity and water quality problems and it recognizes the increasing pressures on water supply to provide for both human and ecosystem needs. The category of *water use* includes not only research questions about managing human consumptive and nonconsumptive use of water, but also about the use of water by aquatic ecosystems and endangered or threatened species. The third category, *water institutions*, emphasizes the need for research into the economic, social, and institutional forces that shape both the availability and use of water.

After review and reconsideration, the committee concluded that the priorities enumerated in the *Envisioning* report constitute the most comprehensive and current best statement of water resources research needs. Moreover, successful pursuit of that research agenda could provide answers to the central questions posed in Chapter 1. However, the list of research topics is not ranked, either within the three general categories or as a complete set of 43. An absolute ranking would be difficult to achieve, as all are important parts of a national water resources research agenda. Furthermore, the list of research priorities can be expected to change over time, reflecting both changes in the generators of such lists and in the conditions to which they are responding. This chapter, thus, provides a mechanism for reviewing, updating, and prioritizing research areas in this and subsequent lists. It should be noted that the 43 research areas in Table 3-1 are of varying complexity and breadth. In addition, the committee expanded research area #21 (develop more efficient water use) from the version found in the *Envisioning* report to include all sectors rather than just the agricultural sector.

The increasing urgency of water-related issues has stimulated a number of scientific societies and governmental entities, in addition to the WSTB, to produce

TABLE 3-1 Water Resources Research Areas that Should Be Emphasized in the Next 10–15 Years

Water Availability

1. Develop new and innovative supply enhancing technologies
2. Improve existing supply enhancing technologies such as wastewater treatment, desalting, and groundwater banking
3. Increase safety of wastewater treated for reuse as drinking water
4. Develop innovative techniques for preventing pollution
5. Understand physical, chemical, and microbial contaminant fate and transport
6. Control nonpoint source pollutants
7. Understand impact of land use changes and best management practices on pollutant loading to waters
8. Understand impact of contaminants on ecosystem services, biotic indices, and higher organisms
9. Understand assimilation capacity of the environment and time course of recovery following contamination
10. Improve integrity of drinking water distribution systems
11. Improve scientific bases for risk assessment and risk management with regard to water quality
12. Understand national hydrologic measurement needs and develop a program that will provide these measurements
13. Develop new techniques for measuring water flows and water quality, including remote sensing and *in situ*.
14. Develop data collection and distribution in near real time for improved forecasting and water resources operations
15. Improve forecasting the hydrological water cycle over a range of time scales and on a regional basis
16. Understand and predict the frequency and cause of severe weather (floods and droughts)
17. Understand recent increases in damages from floods and droughts
18. Understand global change and its hydrologic impacts

Water Use

19. Understand determinants of water use in the agricultural, domestic, commercial, public, and industrial sectors
20. Understand relationships between agricultural water use and climate, crop type, and water application rates
21. In all sectors, develop more efficient water use and optimize the economic return for the water used
22. Develop improved crop varieties for use in dryland agriculture
23. Understand water-related aspects of the sustainability of irrigated agriculture
24. Understand behavior of aquatic ecosystems in a broad, systematic context, including their water requirements
25. Enhance and restore of species diversity in aquatic ecosystems
26. Improve manipulation of water quality and quantity parameters to maintain and enhance aquatic habitats
27. Understand interrelationship between aquatic and terrestrial ecosystems to support watershed management

Water Institutions

28. Develop legal regimes that promote groundwater management and conjunctive use of surface water and groundwater

continued

TABLE 3-1 Continued

29. Understand issues related to the governance of water where it has common pool and public good attributes
30. Understand uncertainties attending to Native American water rights and other federal reserved rights
31. Improve equity in existing water management laws
32. Conduct comparative studies of water laws and institutions
33. Develop adaptive management
34. Develop new methods for estimating the value of nonmarketed attributes of water resources
35. Explore use of economic institutions to protect common pool and pure public good values related to water resources
36. Develop efficient markets and market-like arrangements for water
37. Understand role of prices, pricing structures, and the price elasticity of water demand
38. Understand role of the private sector in achieving efficient provision of water and wastewater services
39. Understand key factors that affect water-related risk communication and decision processes
40. Understand user-organized institutions for water distribution, such as cooperatives, special districts, and mutual companies
41. Develop different processes for obtaining stakeholder input in forming water policies and plans
42. Understand cultural and ethical factors associated with water use
43. Conduct *ex post* research to evaluate the strengths and weaknesses of past water policies and projects

SOURCE: Adapted from NRC (2001a), which identifies the researchable questions associated with each topic.

their own lists of research priorities. For example, the American Society of Limnology and Oceanography recently convened a workshop to draft a list of emerging research issues (ASLO, 2003). These issues included the biogeochemistry of aquatic ecosystems, the influence of hydrogeomorphic setting on aquatic systems, the impacts of global changes in climate and element cycles, and emerging measurement technologies. This list builds on the comprehensive analysis of research priorities for freshwater ecosystems set forth in *The Freshwater Imperative* (Box 2-1; see also Naiman et al., 1995). Another list of research priorities was recently assembled by the European Commission (2003), Task Force Environment–Water, which emphasizes water availability and water quality and the social, economic, and political aspects of water management. Like the NRC (2001a) report, this research agenda sets forth broad areas of research, with more specific "action lines" within high-priority areas. However, the approach differs from NRC (2001a) in that water quality is separated from water availability, and the socioeconomic and political research agenda is oriented toward crisis management. The U.S. Global Change Program also identified interrelated issues of quantity, quality, and human society as key research needs (Gleick et al., 2000);

this research agenda emphasizes the development of models and methods of prediction as well as data collection and monitoring systems, and it emphasizes research on the socioeconomic and legal impacts of climate change.

This brief review of selected contemporary lists of research priorities, as well as the lists of research priorities shown in Box 2-1, illustrates that the articulation and the ranking of research topics vary with the entity charged to develop a research agenda. It can be anticipated that future lists of priorities will also differ from these.

A METHOD FOR SETTING PRIORITIES OF A NATIONAL RESEARCH AGENDA

The business of setting priorities for water resources research needs to be more than a matter of summing up the priorities of the numerous federal agencies, professional associations, and federal committees. Indeed, there is no logical reason why such a list should add up to a nationally relevant set of priorities, as each agency has its own agenda limited by its particular mission, just as each disciplinary group and each committee does. There is a high probability that research priorities not specifically under the aegis of a particular agency or other organization will be significantly neglected. Indeed, the institutional issues that constitute one of the three major themes in Table 3-1 are not explicitly targeted in the mission of any federal agency. This is the current state of affairs in the absence of a more coordinated mechanism for setting a national water resources research agenda.

A more rigorous process for priority setting should be adopted—one that will allow the water resources research enterprise to remain flexible and adaptable to changing conditions and emerging problems. Such a mechanism is also essential to ensure that water resources research needs are considered from a national and long-term perspective. The components of such a priority-setting process are outlined below, in the form of six questions or criteria that can be used to assess individual research areas and thus to assemble a responsive and effective national research agenda. In order to ensure the required flexibility and national-scale perspective, the criteria should also be applied to individual research areas during periodic reviews of the research enterprise.

1. **Is there a federal role in this research area?** This question is important for evaluating the "public good" nature of the water resources research area. A federal role is appropriate in those research areas where the benefits of such research are widely dispersed and do not accrue only to those who fund the research. Furthermore, it is important to consider whether the research area is being or even can be addressed by institutions other than the federal government.

2. **What is the expected value of this research?** This question addresses the importance attached to successful results, either in terms of direct problem solving or advancement of fundamental knowledge of water resources.

3. **To what extent is the research of national significance?** National significance is greatest for research areas (1) that address issues of large-scale concern (for example, because they encompass a region larger than an individual state), (2) that are driven by federal legislation or mandates, and (3) whose benefits accrue to a broad swath of the public (for example, because they address a problem that is common across the nation). Note that while there is overlap between the first and third criteria, research may have public good properties while not being of national significance, and vice versa.

4. **Does the research fill a gap in knowledge?** If the research area fills a knowledge gap, it should clearly be of higher priority than research that is duplicative of other efforts. Furthermore, there are several common underlying themes that, given the expected future complexity of water resources research, should be used to evaluate research areas:

- the **interdisciplinary** nature of the research
- the need for a **broad systems context** in phrasing research questions and pursuing answers
- the incorporation of **uncertainty** concepts and measurements into all aspects of research
- how well the research addresses the role of **adaptation** in human and ecological response to changing water resources

These themes, and their importance in combating emerging water resources problems, are described in detail in this chapter.

5. **How well is this research area progressing?** The adequacy of efforts in a given research area can be evaluated with respect to the following:

- current funding levels and funding trends over time
- whether the research area is part of the agenda of one or more federal agencies
- whether prior investments in this type of research have produced results (i.e., the level of success of this type of research in the past and why new efforts are warranted)

These questions are addressed with respect to the current water resources research portfolio in Chapter 4.

6. **How does the research area complement the overall water resources research portfolio?** The portfolio approach is built on the premise that a diverse mix of holdings is the least risky way to maximize return on investments. When applied to federal research and development, the portfolio concept is invoked to mean a mix between applied research and fundamental research (Eiseman et al., 2002). Indeed, the priority-setting process should be as much dedicated to ensuring an appropriate balance and mix of research efforts as it is to listing specific research topics. In the context of water resources, a diversified portfolio would capture the following desirable elements of a national research agenda:

- multiple national objectives related to increasing water availability, improving water quality and ecological functions, and strengthening institutional and management practices
- short-, intermediate-, and long-term research goals supporting national objectives
- agency-based, contract, and investigator-driven research
- both national and region-specific problems being encompassed
- data collection needs to support all of the above

Thus, the water resources research agenda should be balanced in terms of the time scale of the effort (short-term vs. long-term), the source of the problem statements (investigator-driven vs. problem-driven), the goal of the research (fundamental vs. applied), and the investigators conducting the work (internally vs. externally conducted). An individual research area should be evaluated for its ability to complement existing research priorities with respect to these characteristics. Definitions of these terms are provided in Box 3-1, and the appropriate balance among these categories is addressed in Chapters 4 and 6.

Furthermore, it is important to consider whether the research fills gaps in the desired mix of water availability, water use, and institutional topics (as demarcated in Table 3-1). A final level of evaluation would consider how well the research responds to the four themes described in this chapter (interdisciplinarity, broad systems context, evaluation of uncertainty, and adaptation).

To summarize, a balanced water resources research agenda will include items of national significance for which a federal role is necessary; fill knowledge gaps in all three topical areas (water availability, water use, and institutions); incorporate a mixture of short-term and long-term research, basic and applied investigations, investigator-initiated and mission-driven research, and internal and external efforts; and build upon existing funding and research success. As noted above, some of these issues are addressed in subsequent chapters, with respect to the current water resources research agenda (see Table 3-1). The remainder of this chapter expands upon the four overarching themes that should form the context within which water resources research is conceptualized and performed.

BOX 3-1
Definitions of Research

In order to assess the scope and adequacy of the national research agenda in water resources, it is first necessary to articulate what is meant by "research." Research encompasses intellectual inquiry in pursuit of new knowledge. However, this inquiry can take place across many dimensions of temporal and spatial scale, purpose, and organization. After reviewing the varieties of activities classified as "research" by the federal agencies, the committee developed a taxonomy of research categories that was used to assess the distribution and balance of the national water resources research agenda. Following is a description of the categories as used by the committee to assess the current status of water resources research.

Short-term vs. Long-term

It is important to specify the time scale over which the research is done and over which the results of the research may be applied. "Short-term" research refers to research efforts that are conceptualized and prioritized over a maximum of five-year time frames and conducted over shorter periods of time (two to three years) and that are applicable on immediate time scales. Short-term research is expected to produce immediate results that can be directly applied to current problems. Developing methods of optimizing the use of current water supplies, a research priority of the U.S. Bureau of Reclamation, is a typical example of short-term research. In contrast, "long-term" research refers to research efforts that are conceptualized and prioritized over time frames of more than five years and are usually carried out over relatively long time frames (greater than five years) and/or produce results that will only be applicable to management or further research over similarly long time scales. Examples include the Long-term Ecological Research sites of the National Science Foundation (NSF) and the research watersheds maintained by the U.S. Forest Service, as well as research conducted on fundamental aspects of water science.

Fundamental vs. Applied

Research can be evaluated in terms of the type of knowledge that is sought. Traditionally, research that is solely inspired by curiosity—a quest to understand the world and generate new knowledge—is thought to be "fundamental." Such research is contrasted with "applied research," which is designed to solve a specific, contemporary problem. However, a more

continued

BOX 3-1 Continued

realistic representation of the these categories distinguishes two types of fundamental research, which can be denoted as "pure basic research," which is conducted without respect to any practical application, and "use-based basic research" in which an ultimate application informs research that seeks the basic knowledge necessary to solve a problem (Stokes, 1997). The term "fundamental" is used in this report to encompass those activities intended to generate new knowledge; it includes both that research conducted without respect to any practical application and that inspired by the need for solutions to real-world problems. The term "applied research" is used to encompass those activities that seek to determine if and how current knowledge can be applied to solving problems. This formulation is in accord with the portrayal of research in "Pasteur's quadrant" as a two-dimensional set of continua (Stokes, 1997). In accordance with these definitions, research may be immediately applicable to management problems and yet be "fundamental" if the resolution of those problems involves the production of new understanding of basic phenomena. For example, research contributing to an understanding of groundwater flow in fractured rock aquifers is fundamental research, as this is a poorly understood topic in hydrogeology. However, because there are many fractured rock aquifers that are major water sources for consumptive use and/or are contaminated, the knowledge has immediate application. In contrast, research on the applicability of readily available treatment technologies to remediate contamination in a fractured rock aquifer would be applied research, as it addresses the uses to which existing knowledge may be put.

Investigator-driven vs. Mission-driven

Investigator-driven research is initially conceived by an individual or group of individuals, through imaginative and original thought applied to existing knowledge in a field, and it is conducted as a result of the initiative of the scientist in finding funds to support the research effort. It is sometimes described as curiosity-driven. Such research is usually conducted after external peer review of a research proposal submitted in competition with other investigator-initiated proposals. The research programs of the NSF are the standard for such research. An example might be research exploring a previously unknown mechanism by which a contaminant interferes with cell physiology, which an investigator has thought about and wants to verify experimentally. In contrast, mission-driven research is conducted in response to a problem area identified by and consistent with both an agency mission statement and/or a congressional

continued

BOX 3-1 Continued

mandate in particular legislation. Such problem statements are developed by agency staff and administrators, who then seek out the appropriate mix of scientists to develop a research program to address the problem. While the ingenuity and originality of the scientific approach are highly valued in such research, they do not typically contribute to the initial definition of the scientific problem at hand. An example might be determination of exposure risks for a class of contaminants; the mission is to regulate risk from a class of pollutants, and the goal of the research is to satisfy the performance of this mission.

Internal vs. External

Research can be evaluated in terms of the institutional affiliation of the individuals carrying out the activity. "Internal" research is conducted by investigators employed by the agency funding the work. "External research" is conducted by investigators in institutions other than the funding agency. The large majority of external research is conducted by faculty at institutions of higher education, through grants and contracts with funding agencies.

Overlap Among the Categories

Gradations exist within each category of research, such that a research project may be of, for example, "intermediate term." However, most agency research programs sponsor research that is close enough to one extreme or the other on each scale to be satisfactorily classified by the above typology. This is particularly true for the latter two classifications.

There is considerable overlap among these categories; indeed, in practice they grade into each other, forming continua of research characteristics. Thus, the majority of long-term research is also fundamental research, whereas short-term research is often, but not always, applied. Much of the short-term research is conducted internally, particularly by agencies whose missions are focused on solving current problems. Short-term research is also likely to be mission-driven, for the same reason. Investigator-driven research is, by contrast, most likely to be conducted externally, by individuals based at universities, research institutes, and other nongovernmental organizations, and it is more likely to be fundamental and long-term. Although there are clear correlations among these categories, it is important to note that there is much research being conducted that combines the categories in other ways.

THEMES OF FUTURE WATER RESOURCES RESEARCH

There are several common underlying themes that should be used to (1) integrate and reconcile the numerous lists of research priorities currently being generated by agencies and scientific societies and (2) provide some overall direction to the multiple agencies and academic entities that carry out water resources research. These themes are interdisciplinarity, a broad systems context, uncertainty, and adaptation in human and ecological response to changing water resources.

The term *interdisciplinarity* refers to the fact that no question about water resources can be now adequately addressed within the confines of traditional disciplines. The research community recognizes that the physical, chemical, and biological/ecological characteristics of water resources are causally and mechanistically interrelated, and all are profoundly affected by the human presence in the environment. Therefore, it is necessary to understand water resources with reference to a range of natural and social scientific disciplines.

The phrase *broad system context* refers to the perception that all properties of water are part of a complex network of interacting factors, in which the processes that connect the factors are as important as the factors themselves. Both interdisciplinarity and broad systems context place water resources within the emerging field of complex systems (Holland, 1995; Holland and Grayston, 1998).

Uncertainty—the degree of confidence in the results and conclusions of research—has always been an important component of scientific research. All measurements and observations entail some degree of error, as do methods of data analysis, estimation, and modeling. Understanding the sources and amounts of uncertainty attached to estimates of flow, water quality, and other water resource variables is crucial, because so many practical and often expensive decisions hinge on the results. In short, understanding and measuring uncertainty are central to making informed decisions about water resources. Furthermore, an emphasis on uncertainty also implies attention to the extent and quality of the data available for generating estimates of important variables; this attention in turn implies a need to improve technologies for research and monitoring. Finally, an understanding of the uncertainties in data, models, and scientific knowledge lies at the heart of risk analysis and the development of policies and strategies to handle complex environmental problems (Handmer et al., 2001).

Finally, *adaptation* is a key component of the human, as well as ecological, response to the ever-changing environment. Human society has always changed in response to changing resources; the challenge is now to anticipate environmental changes and develop adaptive responses before catastrophe or conflict force such evolution. This is particularly pressing as research ascertains the impact of human activities on ecosystems, such as greenhouse gas release into the atmosphere and deforestation. Adaptation may involve modifying social mores and norms or forming new government policies including economic policies. For example, there is little doubt among many researchers that emerging water scarcity will demand greatly altered expectations and behaviors in society. It may also involve new methods of managing resources in which flexibility to respond to unanticipated or rapidly occurring problems is the guiding principle.

These four themes are illustrated below, using a subset of the research priorities developed in Table 3-1. The portfolio of existing water resources research tends not to be organized along these thematic lines.

INTERDISCIPLINARY NATURE OF RESEARCH

The need for expertise from many disciplines to solve individual water resource problems is widely recognized and has produced repeated calls for collaborative, interdisciplinary approaches to research (Cullen et al., 1999; Naiman and Turner, 2000; Jackson et al., 2001). For example, aquatic ecosystems research now emphasizes the tight linkages between the traditional biological and ecological issues and both hydrology and human use of water (Poff et al., 1997; Richter et al., 1997). Similarly, the transformations of nutrients and pollutants reflect the interplay of hydrology and microbial ecology (Brunke and Gonser, 1997). Examples of several research areas from Table 3-1 are given below to elaborate on the interdisciplinary nature of water resources research.

➢ Research that addresses the fate and transport of contaminants (#5 in Table 3-1) is necessarily interdisciplinary. Contaminants introduced into surface waters can follow a number of different pathways through the environment depending on water and sediment movement, the domain of hydrologists and geologists, respectively. Some contaminants are adsorbed by soil particles; the rate of adsorption depends on the mineral materials constituting the particles, their organic matter content, and the chemical nature of the contaminant—understanding of which requires the tools of physical chemistry. Both adsorbed and dissolved contaminants may be subject to microbial transformation; the rates of degradation, the microbes capable of such metabolic activity, and the environmental conditions under which their activity is maximized must be determined by microbiologists. The contaminant may be taken up by plant roots, (which is the basis of phytoremediation of hazardous wastes—Pban et al., 1995; Terry and Banuelos, 2000). Consumption of sediment particles by filter-feeding organisms in the waterbody, with subsequent transfer through the food web, can also distribute the contaminant through the ecosystem; ecological analysis of the food web architecture, the purview of ecologists, may help predict biomagnification and impacts on species of concern. The balance of these transport and transformation processes differs for different types of contaminants (e.g., metals, pesticides, chlorinated hydrocarbons, nonaqueous compounds) and complex mixtures of contaminants, and little is known about the extent of these processes for emerging contaminants (e.g., endocrine disrupters and pharmaceuticals). This very brief outline of contaminant fate and transport makes it clear that this research priority necessitates a collaborative effort by physical chemists, soil scientists, hydrologists, geologists, microbiologists, plant scientists, and ecologists.

➢ The research needed to improve manipulation of water quality and quantity parameters to maintain and enhance aquatic habitats (#26 in Table 3-1) must be grounded in the connection between hydrology and the viability of the organisms that make up aquatic ecosystems. Recent research has clearly identified the flow regime as the critical component that defines and structures all types of aquatic ecosystems (NRC, 1992, 2002; Poff et al., 1997; Richter et al., 1997). This includes not only the amounts and flow rates of water within an ecosystem, but its patterns of variation, including extreme depths and flow rates, the frequency of occurrence of extremes, the seasonality and interannual variability of these descriptors, and the importance of fluctuating dry and wet conditions (Poff et al., 1997). How particular flow regimes influence the structure of aquatic ecosystems is mostly unknown, even while the management of these ecosystems is critically dependent on such knowledge. For example, plans for controlled release of water from reservoirs to restore downstream aquatic systems and riparian habitat must be based on a comprehensive understanding of the flow regimes (including the main river channel, the tributary channels, and the floodplain and riparian wetlands) as well as the life requirements of the organisms of concern. This requires the collaboration of hydrologists, river geomorphologists, and sedimentologists with ecologists, ichthyologists, and conservation biologists. Such a multidisciplinary group of scientists was recently involved in controlled releases at Glen Canyon Dam on the Colorado River, with the goal of rebuilding eroding sand bars through sediment scour and subsequent deposition (Webb et al., 1999; Cohn, 2001; Patten and Stevens, 2001; Stevens et al., 2001; Powell, 2002).

Similarly, wetlands are structured by water regimes in which very small variations in flow timing and amounts, in seasonal patterns of flow variation, in flow extremes, and in the duration of wet and dry events have very large effects on the biota (Mitsch and Gosselink, 2000; NRC, 2001b). Withdrawals of both groundwater and surface waters for human use can alter the flow regime, such that even subtle alterations can have large effects on the biota and function of the downgradient wetlands. Current controversy about the failure of mitigation methods and policy to meet the goal of "no net loss" of wetlands (Turner et al., 2001) is rooted in the difficulty of reproducing wetland hydrology in created and restored wetlands (NRC, 1995, 2001b). At the same time, the institutions and policies that are used to implement the goal of "no net loss" are being questioned and challenged. Wetland restoration thus demands research that integrates hydrology, plant and animal ecology, and social science.

➢ The sustainability of irrigated agriculture (#23 in Table 3-1), particularly in arid and semiarid regions, is another example where a multidisciplinary approach is urgently needed. There are numerous factors that can confound the successful operation of irrigation projects on a sustainable basis. Problems related to climate variability, soil salinity, deterioration of the irrigation infrastructure, and social instability contributed to the collapse of the ancient empires, like the Akkadians and Sassanians who lived in the Tigris and Euphrates River valley, or the Hohokams who prospered for a millennium along the Gila and Salt rivers of now south-central Arizona (Postel, 1999). Today's challenges are expected to be similar, because irrigation agriculture is associated with arid and semiarid environments where climate variability significantly impedes the successful long-term operation of these systems. In modern times, storage provided by large dams has reduced the impact of short-term fluctuations in climate. However, the looming prospect of global climate change, coupled with water demands of growing populations, has tremendous implications for irrigated agriculture in the next century (NAST, 2000).

The research challenges are to provide better projections of how climate might change and to improve hydrologic observation systems to document these changes (NAST, 2000). In addition, because large-scale structural solutions for water supply for irrigated agriculture are difficult to justify on social and economic grounds (Pulwarty, 2003), social science research on determinants of water use in the agricultural sector and agronomic research on improved crop varieties for dryland agriculture are needed. The problem of sustaining irrigated agriculture becomes even more interdisciplinary when one considers the need to understand the response of soils and surface water systems (in terms of chemistry and ecology) to alterations in irrigation return flows and the need to understand how economics might produce flexible strategies for irrigation. Assessments like those relating to the restoration of the Colorado River delta (Luecke et al., 1999) or the San Francisco Bay delta (McClurg, 1997) make clear the inherent multidisciplinarity of developing water supply systems for irrigated agriculture within an environment of competing demands and constraints.

➢ The control of nonpoint source pollutants (#6 in Table 3-1) such as fertilizers, pesticides, and animal wastes, most of which emanate from agriculture and urban sources, is another problem in which interdisciplinary research will be essential. Nonpoint source pollutants have been shown to degrade the quality of groundwater and surface waters across the United States (USGS, 1999), and in many cases they are a much larger contributor to poor water quality than are point sources.

Efforts are underway to reduce the nonpoint source contamination of the nation's waters (e.g., Mississippi River Task Force, 2001). However, the enormous scope and scale of the problem are daunting, as land-use practices in several sectors of the economy often result in degradation of water resources in areas far downstream from the site(s) of impact. For example, excessive loading of nitrogen derived mainly from agriculture in the Midwest has contributed to an oxygen-depleted zone in the Gulf of Mexico that can be as large as the state of New Jersey (Goolsby and Battaglin, 2000). Solving this problem requires not only resolving multiple scientific questions, but also resolving social, economic, and political complexities at scales ranging from the local to the national. Combating nonpoint source pollution will require both basic and applied research. For example, although good progress is being made in elucidating factors controlling contaminant loading (e.g., Alexander et al., 2000; Dubrovsky et al., 1998; Porter et al., 2001), more work is required to understand the fate and transport of nonpoint source pollutants and their fundamental effects on human and environmental health, particularly for pesticides and their transformation products (USGS, 1999). This understanding will require decades of high-resolution chemical and biological monitoring coupled with new analytical and modeling approaches.

The key physical approaches for controlling nonpoint source contamination are local mitigation strategies provided by wetlands, sedimentation ponds, and riparian areas along streams, and land-management strategies that reduce runoff and chemical use. Mitigation is an expensive option, both in terms of implementation and reductions in farmed area. Considerable research will be needed in proof-of-concept, design, and in cost/benefit analyses, requiring the participation of ecologists, soil scientists, hydrologists, and geologists to determine the appropriate size, type, and placement of structures. Changes to farming practices on a continental scale will require equally complex research by agronomists, soil scientists, hydrologists, economists, and social scientists because broad stakeholder education and involvement, voluntary actions, new legislative authority, and coordination across localities and regions will be necessary to implement such changes (Mississippi River Task Force, 2001). Finally, contaminant fluxes from land to streams and rivers may well undergo chronic increases as a result of larger rainfall events associated with future climate change. Thus, progress in controlling nonpoint contamination will require interdisciplinary research linking the historically important areas of agriculture, hydrology, and biology with emerging areas of climate change, natural resource economics, education, and human dimensions of decision making.

BROAD SYSTEMS CONTEXT

The systems approach mandates that a problem be addressed by specifying the entities that contribute to the problem, the linkages among these entities, the logical or physical boundaries to the system, and the inputs and outputs to the system as a whole (in other words, linkages to entities deemed to be outside the system). The idea has its roots in physics, in which a "system" is a thermodynamic concept related to the flow and conservation of energy. The linkages among entities within a system are as important as the entities themselves; thus, a system is more than the sum of its parts (see Box 3-2). Systems usually show nonlinear dynamics, and the nonlinearities among sets of linked entities often lead to unanticipated and complex behavior, and also to surprises—events that cannot be exactly predicted, or that are outside the realm of prior experience. Indeed, these characteristics of system behavior have been highlighted as key aspects of environmental problems (NRC, 1997a). Thus, considering water resources research within a broad systems context implies elucidating interrelationships among entities that, at first glance, might not be thought to be related. This approach also mandates that small-scale problems be viewed within a larger-scale perspective, which may profoundly alter the understanding of causal and quantitative relationships.

The need to view some of the research priorities set forth in Table 3-1 within a broad systems context is illustrated below.

➢ Understanding national hydrologic measurement needs (#12 in Table 3-1) will require a systems approach. Human activities can alter water flow regimes in a variety of ways and at a variety of scales. The limits of tolerance of key organisms and species assemblages to changes in flow regime, particularly their tolerance to changes in seasonality, to extremes, and to the frequency and duration of hydrologic changes, are largely unknown. Yet many human-induced alterations to water resources involve these types of changes. It is clear that decisions about the types, frequencies, and spatial distribution of a set of hydrologic measurements can only be made by understanding the broad systems context within which the hydrologic component occurs. This context must include upstream, downstream, regional, and even continental-scale influences on linked hydrologic processes. Similarly, measurements must reflect the competing needs of all potential users of water (both human and ecological), including those far downstream. Consumptive and nonconsumptive water uses and ecosystem water use have different temporal and spatial patterns of demand. Being able to forecast the hydrologic conditions that will affect each of the user groups within a given hydrologic unit will likely require many different types of hydrologic data. By viewing the problem from a broad systems context, strategies for the efficient collection of hydrologic data can be developed.

As an example, the Idaho Department of Water Resources increasingly must resolve conflicts among citizens concerning competing demands for (and assertion of rights over) surface water and groundwater, and it also must resolve interstate water conflicts between Idaho and neighboring states (Dreher, 2003). Provision of adequate water for the habitats of endangered and threatened aquatic species is also part of the state's responsibilities. Idaho contains six aquifers that span interstate lines and that affect surface water flows in adjoining states. Currently, management of both groundwater and surface water supplies is being undertaken without adequate knowledge of the connections between the two sources, leading to conflicts and shortages. The lack of a comprehensive understanding of the entire regional hydrogeologic system and its links to both human use and natural ecosystems is leading to increased litigation, with current needs not being met. In order to help resolve these conflicts, management agencies need

BOX 3-2
The Use of a Broad Systems Context in Addressing Water Research

The traditional approach to problem solving suggests that a dependent variable results from the action of one or a small number of independent variables. An illustration of the traditional approach to a water quality problem would be phosphorus pollution of a waterbody thought of primarily with respect to the major sources (wastewater inputs and natural sources).

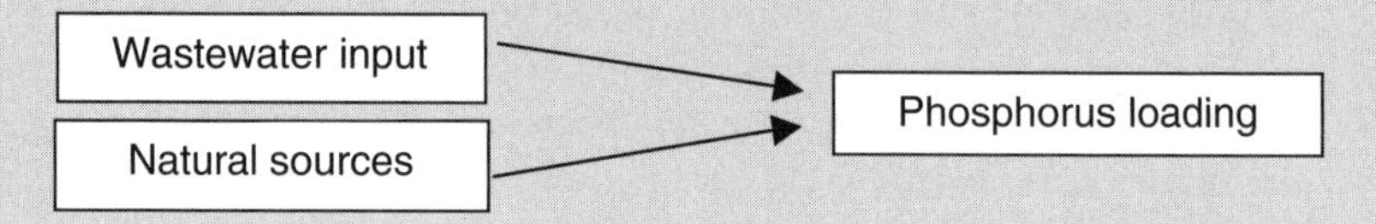

A systems approach, in contrast, emphasizes the fluxes and quantitative relationships among entities within a bounded region. Consider the same problem of phosphorus pollution in a lake, but from a broad systems perspective (see diagram to the right, in which solid arrows indicate phosphorus flows, and dashed arrows indicate other effects). The phosphorus content of the lake is the result of input and output fluxes to compartments within the lake—i.e., algal uptake and release through the decomposition of dead algal cells; deposition to and mobilization from the sediments; inputs from outside the lake such as flowing water, litter fall, and bedrock sources; and outflows from the system. The diagram

accurate measurements of water flows and water stocks over a range of temporal and spatial scales. Moreover, the influences of natural processes, natural climate variability, and human intervention in the water system must be monitored.

➢ A broad systems context is essential to understanding the hydrologic impacts of global change (#18 in Table 3-1). Anticipated changes in temperature regimes and temperature extremes will affect all components of the hydrologic cycle, and numerous feedbacks between the hydrologic cycle and temperature regime will occur. Moreover, changing temperature and precipitation patterns will affect nonaquatic ecosystems, such as upland forests and savannas, in ways that will feed back on hydrology (Raupach, 1999; Valentini et al., 1999). For example, climate changes that alter the extent and density of forest cover will affect both also indicates that there are feedbacks; for example, fisheries represent both a net flux of phosphorus out of the lake, but they also affect the regional economy, which in turn can affect both the flow of wastewater and nonpoint sources pollutants into the lake as well as water flow rates into the lake via upstream diversions of water. Thus, understanding phosphorus pollution in the lake depends on understanding the nature of fluxes and feedbacks among the components of the system and the factors controlling each flux, as well as understanding the components themselves.

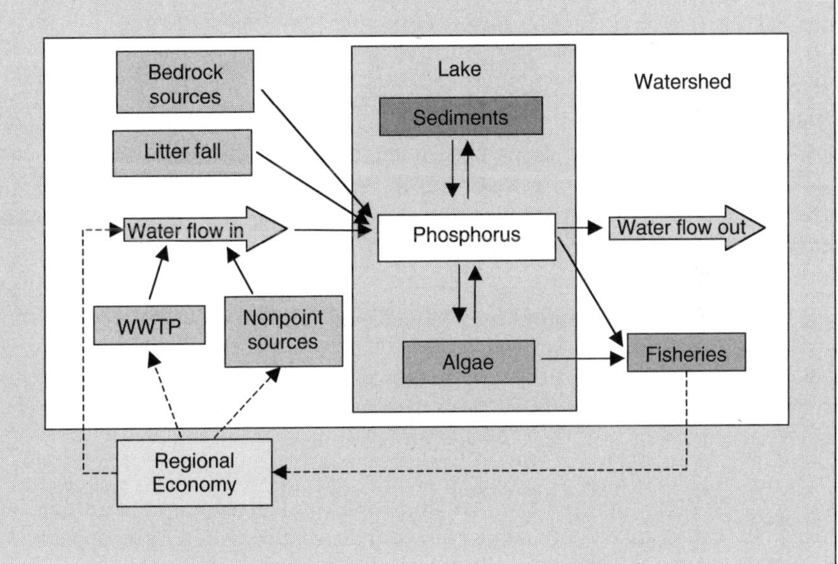

transpiration rates from vegetation and evaporation rates from the soil surface, thus altering soil and atmospheric moisture content and the likelihood of rain and forest fire. These in turn will have large effects on regional hydrology. These connections, which have been well documented for tropical rain forests, are germane to understanding the connections between hydrology and climate worldwide.

Moreover, the driving force for global climate change—the rise in greenhouse gas concentrations associated with human activities—will also affect aquatic ecosystems in ways that may amplify or dampen the effects of hydrologic change alone. For example, higher CO_2 concentrations will alter leaf chemistry and the relative growth rates of different plant species. Both changes may affect the palatability of litter to decomposer and consumer organisms, in turn affecting decomposition rates, nutrient cycling rates, and ultimately the density and species composition of the plant community. Changing CO_2 concentrations may also affect pH of the water, with cascading effects on the biota, although changes in flow regime may interact with increased dissolution of CO_2 to modify this effect. These feedbacks are being incorporated into the models that are used to predict the effects of greenhouse gas emissions on climate and water resources. Unfortunately, the great complexity of the system results in model predictions that span a range of values too large and uncertain to be usable for regional or local water resource management at this time (Chase et al., 2003).

➢ Understanding determinants of water use in agriculture, domestic, commercial, public, and industrial sectors (#19 in Table 3-1) requires understanding the connections between energy and water in a broad systems context. Nearly all aspects of water supply and use are highly energy-intensive. In the western United States, the ability to supply water to a growing population will depend not just on the availability and cost of water, but also on the availability and cost of energy to move water to population centers and to operate treatment, distribution, and wastewater collection systems. In California alone, water pumping and treatment account for 6.5 percent of total electricity used in the state, or about 15,000 gigawatt-hours per year (California Energy Commission, 2003). In addition, recent power crises in California and Nevada have demonstrated the limitations of the region's energy supplies. Because water and wastewater systems typically exhibit increasing economies of scale, energy may become more of a limiting factor in supplying water to urban areas of the West than water itself. Moreover, the conveyance of large volumes of water has already proved to have profound effects on water quality and the functioning of natural systems. In virtually every major western basin where large-scale water works have been constructed to provide water for irrigation, hydropower, and municipal and industrial uses, natural systems have been disrupted and multiple aquatic species threatened or endangered. As a result, the modern mission of the U.S. Bureau of Reclamation has been transformed from building western water systems to mitigating unintended physical and ecological consequences of these systems. These connections between regional economies, energy supply and economics, regional aquatic ecology, and water supply constitute a complex system.

Just as energy supply interacts with water use in multiple ways, as described above, energy extraction (for example, oil and gas development in the West) similarly affects water use in complex ways. Impacts of energy extraction on biotic resources may affect water supply and water use indirectly, by limiting potential options to manage water resources. For example, recent and rapid development of methane gas resources in the Powder River Basin is causing major disruptions in groundwater supply sources (BLM, 2003). Depending on the method of energy extraction, water quality is often impaired. Drilling muds, for example, frequently contain additives that have the potential to contaminate downstream or downgradient water supplies (EPA, 2000).

UNCERTAINTY

Water resource management relies on monitoring data, scientific understanding of processes in the water cycle and the ecology of aquatic ecosystems, and ultimately predictive models that can forecast hydrologic conditions and biotic and human responses. All of these types of information are subject to uncertainty. Uncertainty results from many sources, including measurement systems that are not sufficiently precise or that do not generate sufficient quantities of high-quality data, instrument failures, human errors in designing and implementing studies, and simply a lack of understanding of the processes and phenomena under investigation. Uncertainty affects both the analysis of data and the construction of models to make water resource predictions. Although inherent to research, uncertainty can be managed by explicit recognition of its occurrence coupled with quantitative methods of measuring its importance and incorporating it into decision making. By describing the degree of uncertainty in research results (and by inference the reliability of the measurements and models), researchers can adjust the expectations for the use of their data and models accordingly. Reliable estimates of uncertainty contribute directly to successful risk management and the development of environmental policy (Funtowicz and Ravetz, 1990; Dovers et al., 2001). It should be noted that the above definition of uncertainty is broader than that espoused by some federal agencies (e.g., the U.S. Army Corps of Engineers, for which uncertainty refers to situations in which the probability of potential outcomes and their results cannot be described by objectively known probability distributions). Below are examples illustrating the importance of the quantification of uncertainty for some of the research priorities listed in Table 3-1.

➢ Our ability to improve hydrologic forecasting over a range of time scales and on a regional basis (#15 in Table 3-1) will depend on developing new methods to quantify and reduce uncertainty in the predictive models used to produce forecasts. There are several climate models, run at major centers around the world, that provide global seasonal and longer-term forecasts at the spatial scale of 60,000 km^2 or coarser. These models generally can produce accurate forecasts of seasonal climate conditions over certain portions of the globe (e.g., Shukla et al., 2000; Goddard et al., 2001). However, they are very sensitive to the initial conditions used to parameterize the models and to the accuracy of data used for applying the models to smaller spatial scales. Thus, the models produce significant errors in both the projected mean values of climatic variables and in the estimates of their variability and extremes when used for hydrologic forecasting (e.g., Strauss, 1993; Risbey and Stone, 1996; Mason et al., 1999; Anderson et al., 1999; Wang and Zwiers, 1999; Kharin and Zwiers, 2000). Use of these forecasting tools must clearly be tempered by quantitative estimates of the uncertainty of the predicted conditions. This is especially the case for developing regionally relevant predictions and incorporating climate forecast models into water resource management tools (Georgakakos and Krzysztofowicz, 2001).

➢ Our ability to understand the assimilation capacity of the environment (#9 in Table 3-1) is predicated on the construction of models that describe the fate, transport, and effects of contaminants. However, as with all models, success is contingent on an understanding of the physical, chemical, and biological processes involved in contaminant dynamics, which have varying degrees of uncertainty. This is strikingly illustrated by research in support of creating a high-level radioactive waste repository. The disposal of high-level nuclear waste presents a unique challenge for the water resources research community because of the waste's extraordinary longevity. The Department of Energy (DOE) has proposed locating a repository at Yucca Mountain, Nevada, and is preparing to submit a license application to the Nuclear Regulatory Commission by the end of 2004.[1] Approximately 70,000 metric tons heavy metal (MTHM) of spent fuel and high-level waste are destined for Yucca Mountain if construction of the repository is approved. Without any natural or engineered barriers, bare waste could result in a peak mean annual dose of about 2.7×10^{10} millirem per year (Saulnier, 2002). The DOE has proposed a repository design intended to reduce this dose to below 10^{-1} millirem per year for a regulatory period of 10,000 years (DOE, 2002).

To predict the fate and transport of contaminants from the proposed repository, the DOE has developed a complex mathematical model called Total System Performance Assessment (TSPA) that itself depends on the output of dozens of process-oriented models. The success of the DOE's license application depends in large measure on the confidence placed in the TSPA predictions of contaminant transport and the technical basis for those predictions. Conceptual and model uncertainty and the explicit quantification of this uncertainty are central to the question of technical basis. As noted by the U.S. Nuclear Waste Technical Review Board in a letter to Congress (NWTRB, 2002): "Resolving all uncertainty is neither necessary nor possible. However, uncertainties about the performance of those components of the repository system relied upon to isolate waste are very important, and information on the extent of uncertainty and assumed conservatism associated with the performance of these components may be important to policy makers, the technical community, and the public." Regardless of policymakers' and the public's varying levels of tolerance for uncertainty, it can still be said that results of research to quantify, and perhaps further reduce, uncertainties can contribute to the quality and credibility of impending public policy decisions.

➢ There is an important role for uncertainty analysis in better understanding the impact of land use changes and best management practices on pollutant loading to waters (#7 in Table 3-1). This research priority is directly related to the development of Total Maximum Daily Loads (TMDLs) and subsequent efforts to

---

[1]This committee was not constituted to determine the merits of Yucca Mountain project or the ripeness of the decision to license a repository there.

remediate polluted waterbodies. Mandated by the Clean Water Act, a TMDL is a calculation of the maximum pollutant loading that a waterbody can sustain and still meet its water quality standards. If the current loadings are higher, then the TMDL must be accompanied by a remedial plan on how to reduce the loadings via best management practices (BMPs). TMDLs are established for an impaired waterbody by using a combination of fate and transport models for the target pollutant or stressor and available waterbody data. This requires both watershed models (which take into account such processes as the movement of pollutants across land) and water quality models (which incorporate in-lake pollutant transport and transformation). Models are also potentially needed to predict the effectiveness of certain BMPs. Many of the watershed and water quality models in use suffer from inadequate representation of physicochemical processes, inappropriate applicability, and lack of training of model users (EPA, 2002). Similarly, the data on which TMDLs are based may be inconsistent in quality or inappropriate in terms of the frequency and extent of sampling. Finally, the methods used to identify impaired waterbodies are often inadequate because of deficiencies in state monitoring networks. All of these problems generate uncertainties in the applicability and effectiveness of the resulting TMDL. The development of improved methods of quantifying uncertainty in both the models and the listing criteria, especially in setting "margin of safety" criteria, is critical if informed decisions about restoring polluted waterbodies are to be made. Indeed, the central role of uncertainty has been a major conclusion of several recent studies critically examining the TMDL program (NRC, 2001c; Borsuk et al., 2002; EPA, 2002).

ADAPTATION

Water resource managers are subject to increasingly diverse, often conflicting forces. For example, it was relatively simple to develop the knowledge base needed to provide predictable amounts of water to agriculture when this was the only use for a water supply. It becomes much more complicated when agricultural uses need to be met while new demands come from urbanizing areas and from governmental and nongovernmental entities demanding water for endangered species or aquatic ecosystem support, such that the total demand exceeds the readily available supply. In such contexts, adaptability becomes essential. Managers, users, and advocates need to have the flexibility to imagine and adopt novel solutions to water resource problems, and researchers in their search for solutions need to have the flexibility to adapt their research to problems that may have been unimaginable in the recent past. Furthermore, the complexity of current problems may demand that combinations of solutions be applied creatively to different components of a problem. This emphasis on adaptability of both the research community and the managers and users of water needs to be an organizing concept for water resources research. Thus, "adaptation" is defined as a combination of flexibility in solving problems and, more fundamentally, a shift in norms and standards that can result from confronting novel situations. A related concept in water resources is that of adaptive management, a learning-while-doing process in which a management action is viewed as an experiment, and as managers learn from their successes and failures, they adjust their management actions accordingly (Holling, 1978; Geldof, 1995; Haney and Power, 1996; Wieringa and Morton, 1996; Lee, 1999; NRC, 1999, 2003b, 2004b).

Below are examples of how adaptation is a key element in addressing some of the research priorities listed in Table 3-1.

➢ Improving the integrity of drinking water distribution systems (#10 in Table 3-1) will have to come at least partially from research that addresses the nation's aging water delivery infrastructure, particularly in the eastern United States (Davies et al., 1997; Levin et al., 2002). It is well known that in-line infiltration into cracked or otherwise compromised water delivery pipes occurs during cases of extreme hydrologic events or even under normal operation when there is transient negative pressure in the pipeline (Besner et al., 2001). During such events, contaminants from the surrounding soil are drawn into the water delivery system. Although replacement of distribution systems can prevent such occurrences, it is not yet known what materials are best for long-term replacement of the systems (McNeill and Edwards, 2001). Upgrading water supply infrastructure is not likely to occur in the near future for many systems for financial reasons (see GAO, 2002). Other options for improving the integrity of drinking water systems, such as better treatment to potable water standards of all water delivered to homes and businesses, is becoming increasingly costly as well. In addition, completely reliable transportation of microbially safe water over long distances cannot always be performed cost-effectively.

This combination of challenges will require adaptability on the part of both researchers and users. For example, creative water delivery systems, such as in-home gray water recycling or dual-home distribution systems (Wilchfort and Lund, 1997) that bring potable water to a few taps and slightly less pure water to other taps for cleaning purposes or industrial needs, will require research. This includes research to develop the technologies to implement such systems and research to understand how people adapt to new modes of obtaining and using water (see Box 3-3) and how such a transition might be effected. Individuals' views of water-related risks (Loewenstein et al., 2001), in-home uses of water, and the value of water resources (Aini et al., 2001) will also need to adapt in order for these technological changes to be successful in maintaining drinking water quality.

➢ The task of enhancing and restoring aquatic ecosystems (#25 in Table 3-1) requires the integration of human and ecological uses of water, a daunting task that will require adaptation on the part of all concerned. As discussed above, natural variability in flow regime and hydroperiod acts to maintain a healthy and

BOX 3-3
Research on Changing Human Perceptions of Water

A comprehensive, coordinated research strategy focused on human beliefs, values, and decision making about water is needed better understand humans' potential to adapt to a changing water environment. In the past 20 years, research has been conducted on people's perceptions of environmental issues (e.g., Slovic, 2000), but little has been done on water specifically. The body of knowledge concerning the factors that affect populations' perceptions of water (Anadu and Harding, 2000), its value (NRC, 1997b; National Water Research Institute, 1999; Aini et al., 2001), its quality (NRC, 2001d; Williams and Florez, 2002), related risks (Lowenstein et al., 2001), and decision processes (Krewski et al., 1995) is not well developed. As an example, limited research has been conducted on the social and political complexity of water reuse as part of a sustainable community (e.g., see Hartley, 2003), and broad issues about public perception and acceptance of reuse remain unaddressed. In addition, research on effective means of communicating water-related risks has received limited attention (e.g., Griffin et al., 1998; Harding and Anadu, 2000; Burger et al., 2001; Parkin et al., 2003).

Only fragmented information is currently available to address water-related issues on the personal, social, or cultural scale. It is known that cultural biases and lifestyle preferences are powerful predictors of risk perceptions (Dake and Wildavsky, 1991). McDaniels et al. (1997) found that a small set of underlying factors (ecological impact, human benefits, controllability, and knowledge) affect lay people's judgments about risks to water resources. One study in the United States indicates that people choose their source of water based on their awareness of water problems, their beliefs that such problems affect them personally, and the duration of the problems (Anadu and Harding, 2000). A much earlier study on water reuse in California indicated that the public favored options that protected public health, enhanced the environment, and conserved scarce water resources (Crook and Bruvold, 1980). In the Southwest, Caucasians and Mexican Americans have been found to have important differences in their views of water quality-related risks, equity, trust, and participation in civic affairs (Williams and Florez, 2002). In the United Kingdom, people's perceptions of power and authority and beliefs in the efficacy of collective action were found to be associated with public views about recreational water (Langford et al., 2000). A study in Canada suggests that people believe that environmental quality (including water quality) is getting worse; they will not support decisions they feel will continue that trend or compromise their health, even if the economy improves (Krewski et al., 1995).

These studies have contributed to knowledge about water-related perceptions and decision processes, but the data are insufficient to provide a complete understanding of the factors that influence individual's decisions about water.

diverse biological community within aquatic and riparian ecosystems. However, human actions to minimize floods and droughts and to provide reliable water for consumption at constant rates can eliminate this natural variability (Dynesius and Nilsson, 1994). In order to balance these effects, management of the water, the ecosystem, and the affected social groups must be adaptive in several respects.

For example, ecological restoration, while guided by ideals of the undisturbed or historical state of the ecosystem, increasingly must accept the lesser but still critical goal of repairing damaged systems to a partially restored state. This will be necessary because of insufficient knowledge of the undisturbed state, permanent alteration of the landscape through built structures and intensive land use, and the prevalence of nearly ineradicable nonnative species. An example is provided by the Laurentian Great Lakes, where overfishing and the onslaught of the sea lamprey brought about the decline of native fishes, including the lake trout. At the same time, exotic species of smaller "forage" fish proliferated, resulting in the famous die-off of alewives that littered Chicago's beaches in the early 1970s. Fisheries managers attempted a bold experiment, importing coho and king salmon from the Pacific Northwest, a highly successful adaptation to a "collapsing" ecosystem. Now with well over one hundred nonnative species, the Great Lakes pose a continuing challenge to ecologists and fisheries managers seeking to manage and restore the ecosystem.

Adaptation is anticipated to be particularly difficult but absolutely essential in large aquatic ecosystems where there are multiple competing interests (fisheries scientists, communities relying on fishing, farmers, water resource and dam managers, etc.) (Peterson, 2000). The scale of conflicts arising from the plexus of interests involved in large-scale ecosystem restoration is illustrated by the recent Klamath (NRC, 2003a) and Columbia River controversies (Gregory et al., 2002; NRC, 1996, 2004a). Clearly, research is needed to develop adaptive approaches to both managing the resources (water, fish, etc.) as well as the various human populations involved in these issues. Flexibility, an understanding that a variety of alternative strategies are possible, and a willingness to adjust previously assumed "rights" will be essential in finding compromises between competing human and ecosystem demands. In addition, the use of adaptive management procedures will be necessary.

➢ The need to understand governance of water (#29 in Table 3-1) and improve equity[2] in current water law (#31 in Table 3-1) is predicated on an awareness of the importance of flexibility or the ability to adapt to new situations. Laws are inherently conservative since their function is to fix in-place rules governing human actions. Generally, certainty and clarity are important objectives of law so

[2]Equity in this context refers to fairness. Equity or fairness is not a scientific concept but is of pivotal importance in jurisprudence and policy making.

that people know what is expected or required and can act in accordance. Thus, for example, investments can be made with the expectation that changes in law will not undo the hoped-for return that motivated the investment. Actions can be taken without fear that a change in the rules will punish the actor. A stable legal system is important economically and socially.

However, this societal interest in stability may conflict with other emerging societal interests in periods of active change. During the 1970s, for example, Congress imposed far-reaching new legal requirements on those whose activities generated certain types of pollution from readily identifiable (point) sources, forcing massive investment in technologically advanced systems for the treatment of particular pollutants prior to their discharge into the environment. The years immediately following enactment of these laws were ones of considerable turmoil and conflict as uncertainties respecting their implementation were disputed and resolved. With these requirements now firmly embedded into the plans and actions of the regulated community, stability has returned. So too has resistance to any significant change in approach, even if such change might better accomplish the objectives of these laws.

Laws governing human uses of water have traditionally been concerned with determining who may make use of the resource and under what conditions. In those states east of the 100th meridian, owners of land adjacent to waterbodies essentially share the ability to use the water (riparian doctrine). Uses must be "reasonable," with reasonable use generally being measured by the harm that might be caused to other riparian users. In the western states, uses are established through a process of appropriation of water—that is, establishing physical control—and then applying the water to a "beneficial use." It is a priority system, protecting full use of available water by those first to appropriate it.

The appropriation system arose in the context of water-scarce settings. Direct use of water from streams initially for mining and then for agriculture was essential, and it required the investment of time and money to build the structures that would make that use possible. Users wanted certainty about their rights of use versus other subsequent users, and the prior appropriation system provided that certainty. The appropriation system does not, however, readily accommodate changing uses of water or integrate new uses. Nor does it incorporate the use of water for serving physical and ecological functions within the hydrologic cycle. This suggests that water laws need to be more adaptable if they are to meet changing societal needs. As a first effort, many western states have adopted water transfer laws to accommodate changing water uses, including environmental needs such as instream flows. These states have successfully combined the certainty of the prior appropriation system with the ability to meet emerging demands.

The process of restoring a sustainable level of physical and ecological integrity to our hydrologic systems must work within long-established legal and institutional structures whose purpose has been to promote and support direct human uses. The challenge is to develop societally acceptable approaches that allow those uses to continue but in a manner that is compatible with ecosystem functionality.

LIMITATIONS TO THE CURRENT WATER RESOURCES RESEARCH ENTERPRISE

The articulation of these four themes—interdisciplinarity, broad systems context, uncertainty, and adaptation—is intended to reorient the disparate research agendas of individual agencies as well as individual researchers. The hope is that an emphasis on these overarching themes will lower barriers to research on newly emerging water resources problems. Research agendas of the federal agencies are driven by their specific mandates, such as the agricultural impacts on water (U.S. Department of Agriculture), water as a component of climate (National Oceanic and Atmospheric Administration), or reservoir management (U.S. Bureau of Reclamation). Often there is a need for agencies to center their missions around clearly articulated, politically prominent issues in order to secure funding. These tendencies promote more narrowly focused research and present barriers to addressing difficult, large-scale problems. Furthermore, agencies are locked into policies devolving from their legislative and administrative history, and they cannot create new policies that cut across administrative or management units; thus, research is constrained by policies that easily become antiquated or irrelevant (Stakhiv, 2003). Finally, water resource problems are frequently conceived to match short-term funding cycles (Parks, 2003), resulting in inadequate knowledge for effective water management.

Similarly, individual scientists frame research in terms of their disciplinary training and work environment, which creates barriers to the kind of research needed to solve the complex problems that are now prominent. Indeed, the reluctance of scientists to reach outside their disciplines has been identified elsewhere as a barrier to effective water resources research (Parks, 2003). Institutional and professional constraints on priority setting also mitigate against effective research because they inhibit creative, innovative, and rapid responses to newly emerging or unanticipated problems.

Water resource problems are commonly assumed to be only local or regional in scope because water management entities and water supply systems operate on these scales. However, some water-related problems have become truly national in scope, either because of their very large spatial scale (e.g., the connection of the upper Mississippi drainage basin with hypoxia in the Gulf of Mexico) or because controversies rage over the same water issues in many states throughout the nation. Unfortunately, the current organization of water resources research promotes site- and problem-specific research, which results in narrowly conceived solutions that are often not applicable to large-scale, complex problems or to similar issues in other regions of the country (Stakhiv, 2003). Federal agencies may see only the local character of a problem, without understanding the sometimes subtle ways in which local problems are widely replicated around the country, and may conclude that such problems are not appropriately addressed with federal resources. State representatives advised the committee that they rarely have the financial or scientific resources to address problems that have local manifestations but national significance. Thus, such research can fail to be carried out because of limitations at both the federal and state levels.

Finally, the ability to carry out research on water resources may be limited by the availability of adequate long-term data (as discussed in Chapter 5). Hydrologic processes are characterized by the frequency with which events of a given magnitude and duration occur. Infrequent but large-magnitude events (floods, droughts) have very large economic, social, and ecological impact. Without an adequately long record of monitoring data, it is difficult, if not impossible, to understand, model, and predict such events and their effects.

By emphasizing interdisciplinarity, broad systems context, uncertainty, and adaptation as overarching research guidelines, the specific research agendas of agencies and, hopefully, individual scientists can be made more relevant to emerging problems. A framework of research priorities based on these overarching themes is more likely to promote flexible, adaptive, and timely responses to novel or unexpected problems than research programs constrained by priority lists developed solely with respect to agency missions. The complexity and urgency of water resource problems demand a framework that widens the scope of inquiry of researchers and research managers and forces them to conduct research in novel ways.

CONCLUSIONS AND RECOMMENDATIONS

Although the list of topics in Table 3-1 is our current recommendation concerning the highest priority water resources research areas, this list is expected to change as circumstances and knowledge evolve. Water resource issues change continuously, as new knowledge reveals unforeseen problems, as changes in society generate novel problems, and as changing perceptions by the public reveal issues that were previously unimportant. Periodic reviews and updates to the priority list are needed to ensure that it remains not only current but proactive in directing research toward emerging problems.

An urgent priority for water resources research is the development of a process for regularly reviewing and revising the entire portfolio of research being conducted. Six criteria are recommended for assessing both the scope of the entire water resources research enterprise and also the nature, urgency, and purview of individual research areas. These criteria should ensure that the vast scope of water resources research carried out by the numerous federal and state agencies, nongovernmental organizations, and academic institutions remains focused and effective.

The research agenda should be balanced with respect to time scale, focus, source of problem statement, and source of expertise. Water resources research ranges from long-term and theoretical studies of basic physical, chemical, and biological processes to studies intended to provide rapid solutions to immediate problems. The water resources research enterprise is best served by developing a mechanism for ensuring that there is an appropriate balance among the different types of research, so that both the problems of today and those that will emerge over the next 10–15 years can be effectively addressed.

The context within which research is designed should explicitly reflect the four themes of interdisciplinarity, broad systems context, uncertainty, and adaptation. The current water resources research enterprise is limited by the agency missions, the often narrow disciplinary perspective of scientists, and the lack of a national perspective on perceived local but widely occurring problems. Research patterned after the four themes articulated above could break down these barriers and promise a more fruitful approach to solving the nation's water resource problems.

REFERENCES

Aini, M. S., A. Fakhru'l-Razi, and K. S. Suan. 2001. Water crisis management: satisfaction level, effect and coping of the consumers. Water Resources Management 15(1):31–39.

Alexander, R. A., R. B. Smith, and G. E. Schwartz. 2000. Effect of stream channel size on the delivery of nitrogen to the Gulf of Mexico. Nature 403:758–761.

American Society of Limnology and Oceanography (ASLO). 2003. Emerging Research Issues for Limnology: the Study of Inland Waters. Waco, TX: ASLO.

Anadu, E. C., and A. K. Harding. 2000. Risk perception and bottled water use. Journal of the American Water Works Association 92(11):82–92.

Anderson, J. L., H. van den Dool, A. Barnston, W. Chen, W. Stern, and J. Ploshay. 1999. Present–day capabilities of numerical and statistical models for atmospheric extratropical seasonal simulation and prediction. Bull. Amer. Meteor. Soc. 80:1349–1361.

Besner, M-C., V. Gauthier, B. Barbeau, R. Millette, R. Chapleau, and M. Prevost. 2001. Understanding distribution system water quality. Journal of the American Water Works Association 93(7):101–114.

Borsuk, M. E., C. A. Stowe, and K. H. Reckhow. 2002. Predicting the frequency of water quality standard violations: a probabilistic approach for TMDL development. Environ. Sci. Technol. 36:2109–2115.

Brunke, M., and T. Gonser. 1997. The ecological significance of exchange processes between rivers and groundwater. Freshwater Biology 37:1–33.

Bureau of Land Management (BLM). 2003. Final Environmental Impact Statement South Powder River Basin Coal. December. http://www.wy.blm.gov/nepa/prbcoal–feis/index.htm.

Burger, J., M. Gochfeld, C. W. Powers, L. Waishwell, C. Warren, and B. D. Goldstein. 2001. Science, policy, stakeholders and fish consumption advisories: developing a fish fact sheet for the Savannah River. Environmental Management 27:4:501.

California Energy Commission. 2003. Water Energy Use in California. http://www.energy.ca.gov/pier/–indust/water_industry.html.

Chase, T. N., R. A. Pielke, Sr., and C. Castro. 2003. Are present day climate simulations accurate enough for reliable regional downscaling? Water Resources Update No. 124:26–34.

Cohn, J. 2001. Resurrecting the dammed: a look at Colorado River restoration. BioScience 51:998–1005.

Crook, J., and W. H. Bruvold. 1980. Public Evaluation of Water Reuse Options. OWRT/RU–80/2. Washington, DC: U.S. Department of the Interior, Office of Water Research & Technology.

Cullen, P. W., R. H. Norris, V. H. Resh, T. B. Reynoldson, D. M. Roseberg, and M. T. Barbour. 1999. Collaboration in scientific research: a critical need for freshwater ecology. Freshwater Biology 42:131–142.

Dake, K., and A. Wildavsky. 1991. Individual differences in risk perception and risk-taking preferences. Pp. 15–24 *In* The Analysis, Communication and Perception of Risk. B. J. Garrick and W. C. Gekler (eds.). New York: Plenum Press.

Davies, C., D. L. Fraser, P. C. Hertzler, and R. T. Jones. 1997. USEPA's infrastructure needs survey. Journal of the American Water Works Association 89(12):30–38.

Department of Energy (DOE). 2002. Yucca Mountain Project: Recommendation by the Secretary of Energy Regarding the Suitability of the Yucca Mountain Site for a Repository under the Nuclear Waste Policy Act of 1982. Washington, DC: DOE Office of Civilian Radioactive Waste Management. Pp. 13–15.

Dovers, S. R., T. W. Norton, and J. W. Handmer. 2001. Ignorance, uncertainty and ecology: key themes. Pp. 1–25 *In* Ecology, Uncertainty and Policy: Managing Ecosystems for Sustainability. J. W. Handmer, T. W. Norton, and S. R. Dovers (eds.). Harlow, UK: Prentice Hall.

Dreher, K. 2003. Presentation to the NRC Committee on Assessment of Water Resources Research. January 9, 2003, Tucson, AZ.

Dubrovsky, N. M., C. R. Kratzer, L. R. Brown, J. M. Gronberg, and K. R. Burow. 1998. Water quality in the San Joaquin–Tulare Basins, California, 1992–95. U.S. Geological Survey Circular 1159. 38 p.

Dynesius M., and C. Nilsson. 1994. Fragmentation and flow regulation of river systems in the northern third of the world. Science 266:753–762.

Eiseman, E., K. Koizumi, and D. Fossum. 2002. Federal Investment in R&D. MR–1639.0–OSTP. RAND Science and Technology Policy Institute. Santa Monica, CA: RAND.

Environmental Protection Agency (EPA). 2000. Profile of the Oil and Gas Extraction Industry. EPA/310–R–99–006. Washington, DC: EPA Office of Compliance Sector Notebook Project.

Environmental Protection Agency (EPA). 2002. The Twenty Needs Report: How Research Can Improve the TMDL Program. EPA841–B–02–002. Washington, DC: EPA Office of Water.

European Commission. 2003. http://europa.eu.int/comm/research/tf–wt1.html/#contents.

Funtowicz, S. O., and J. R. Ravetz. 1990. Uncertainty and quality in science for policy. Dordrecht, The Netherlands: Kluwer Academic Publishing.

Geldof, G. D. 1995. Adaptive water management: integrated water management on the edge of chaos. Water Science and Technology 32:7–13.

General Accounting Office (GAO). 2002. Water Infrastructure: Information on Financing, Capital Planning, and Privatization. GAO 02–764. Washington, DC: GAO.

Georgakakos, K. P., and R. Krzysztofowicz, (eds.). 2001. Special issue on probabilistic and ensemble forecasting. Journal of Hydrology 249:1–196.

Gleick, P. H., et al. 2000. Water: the Potential Consequences of Climate Variability and Change for the Water Resources of the United States. The Report of the Water Sector Assessment Team of the National Assessment of the Potential Consequences of Climate Variability and Change for the U.S. Global Change Research Program. Oakland, CA: Pacific Institute for Studies in Development, Environment, and Security.

Goddard, L., S. J. Mason, S. E. Zebiak, C. F. Ropelewski, R. Basher, and M. A. Cane. 2001. Current approaches to seasonal-to-interannual climate predictions. International Journal of Climatology 21:1111–1152.

Goolsby, D. A., and W. A. Battaglin. 2000. Nitrogen in the Mississippi Basin—Estimating Sources and Predicting Flux to the Gulf of Mexico. U.S. Geological Survey Fact Sheet 135–00.

Gregory, S., H. Li, and J. Li. 2002. The conceptual basis for ecological responses to dam removal. BioScience 52:713–723.

Griffin, R. J., S. Dunwoddy, and F. Zabala. 1998. Public reliance on risk communication channels in the wake of a *Cryptosporidium* outbreak. Risk Anal. 18(4):367–376.

Handmer, J. W., T. W. Norton, and S. R. Dovers. 2001. Ecology, Uncertainty and Policy: Managing Ecosystems for Sustainability. London, UK: Pearson Education Ltd.

Haney, A., and R. L. Power. 1996. Adaptive management for sound ecosystem management. Environmental Management 20:879–886.

Harding, A. K., and E. C. Anadu. 2000. Consumer response to public notification. Journal of the American Water Works Association 92(8):32–41.

Hartley, T. W. 2003. Water Reuse: Understanding Public Perception and Participation. 00–PUM–1. Alexandria, VA: Water Environment Research Foundation.

Holland, J. 1995. Hidden Order: How Adaptation Builds Complexity. Reading, MA: Addison–Wesley.

Holland, J., and S. Grayston. 1998. Emergence: from chaos to order. Reading, MA: Addison–Wesley.

Holling, C. S. (ed.) 1978. Adaptive Environmental Assessment and Management. New York: John Wiley and Sons.

Jackson, R. B., S. R. Carpenter, C. N. Dahm, D. M. McNight, R. J. Naiman, S. Postel, and S. W. Running. 2001. Water in a changing world. Ecological Applications 11:1027–1045.

Kharin, V. V., and F. W. Zwiers. 2000. Changes in the extremes in an ensemble of transient climate simulations with a coupled atmosphere–ocean GCM. Journal of Climate 13:3760–3788.

Krewski, S., P. Slovic, S. Bartlett, J. Flynn, and C. K. Mertz. 1995. Health risk perceptions in Canada.II: worldviews, attitudes and opinions. HERA 1(3):231–248.

Langford, G., S. Georgiou, I. J. Bateman, R. J. Day, and R. K. Turner. 2000. Public perception of health risks from polluted coastal bathing waters: a mixed methodological analysis using cultural theory. Risk Anal. 20(5):691–704.

Lee, K. N. 1999. Appraising adaptive management. Conservation Ecology 3(2):3.

Levin, R. B., P. R. Epstein, T. E. Ford, W. Harrington, E. Olson, and E. G. Reichard. 2002. U.S. drinking water challenges in the twenty–first century. Environmental Health Perspectives 110:43–52.

Loewenstein, G. F., E. U. Weber, C. K. Hsee, and N. Welch. 2001. Risk as feelings. Psychological Bulletin 127(2):267–286.

Luecke, D. F., J. Pitt, C. Congdon, E. Glenn, C. Valdés–Casillas, and M. Briggs. 1999. A Delta Once More: Restoring Wetland Habitat in the Colorado River Delta. Report of the Environmental Defense Fund. 49 p.

Mason, S. J., L. Goddard, N. E. Graham, E. Yulaeva, L. Sun, and P. A. Arkin. 1999. The IRI seasonal climate prediction system and the 1997/98 El Niño event. Bull. Amer. Meteor. Soc. 80:1853–1873.

McClurg, S. 1997. Sacramento–San Joaquin River Basin Study. Report to the Western Water Policy Review Advisory Commission. 75 p.

McDaniels, T. L., L. J. Axelrod, N. S. Cavanagh, and P. Slovic. 1997. Perception of ecological risk to water environments. Risk Analysis 17(3):341–352.

McNeill, L. S., and M. Edwards. 2001. Iron pipe corrosion in distribution systems. Journal of the American Water Works Association 93(7):88–100.

Mississippi River Task Force (MRTF). 2001. Action Plan for Reducing, Mitigating, and Controlling Hypoxia in the Northern Gulf of Mexico. Washington, DC: MRTF.

Mitsch, W., and J. G. Gosselink. 2000. Wetlands, 3rd. ed. New York: Van Nostrand Reinhold.

Naiman, R. J., and M. G. Turner. 2000. A future perspective on North America's freshwater ecosystems. Ecological Applications 10:958–970.

Naiman, R. J., J. J. Magnuson, D. M. McKnight, and J. A. Stanford, eds. 1995. The Freshwater Imperative: A Research Agenda. Washington, DC: Island Press.

National Assessment Synthesis Team (NAST). 2000. Climate Change Impacts on the United States: The Potential Consequences of Climate Variability and Change. U.S. Global Change Research Program. Cambridge, MA: Cambridge University Press. 154 p.

National Research Council (NRC). 1992. Restoration of Aquatic Ecosystems. Washington, DC: National Academy Press.

National Research Council (NRC). 1995. Wetlands: Characteristics and Boundaries. Washington, DC: National Academy Press.

National Research Council (NRC). 1996. Upstream: Salmon and Society in the Pacific Northwest. Washington, DC: National Academy Press.

National Research Council (NRC). 1997a. Building a Foundation for Sound Environmental Decisions. Washington, DC: National Academy Press.

National Research Council (NRC). 1997b. Valuing Ground Water: Economic Concepts and Approaches. Washington, DC: National Academy Press.

National Research Council (NRC). 1999. Downstream: Adaptive Management of Glen Canyon Dam and the Colorado River Ecosystem. Washington, DC: National Academy Press.

National Research Council (NRC). 2001a. Envisioning the Agenda for Water Resources Research in the Twenty–First Century. Washington, DC: National Academy Press.

National Research Council (NRC). 2001b. Compensating for Wetland Losses Under the Clean Water Act. Washington, DC: National Academy Press.

National Research Council (NRC). 2001c. Assessing the TMDL Approach to Water Quality Management. Washington, DC: National Academy Press.

National Research Council (NRC). 2001d. Classifying Drinking Water Contaminants for Regulatory Consideration. Washington, DC: National Academy Press.

National Research Council (NRC). 2002. Riparian Areas: Functions and Strategies for Management. Washington, DC: National Academy Press.

National Research Council (NRC). 2003a. Endangered and Threatened Fishes in the Klamath River Basin: Causes of Decline and Strategies for Recovery. Washington, DC: The National Academies Press.

National Research Council (NRC). 2003b. Adaptive Monitoring and Assessment for the Comprehensive Everglades Restoration Plan. Washington, DC: The National Academies Press.

National Research Council (NRC). 2004a. Managing the Columbia River: Instream Flows, Water Withdrawals, and Salmon Survival. Washington, DC: The National Academies Press.

National Research Council (NRC). 2004b. Adaptive Management for Water Resources Project Planning. Washington, DC: The National Academies Press.

Nuclear Waste Technical Review Board (NWTRB). 2002. Letter Report to Congress and the Secretary of Energy, January 24, 2002. Arlington, VA: NWTRB. http://www.nwtrb.gov/-reports/2002ltr.pdf.

National Water Research Institute. 1999. The Value of Water. Fountain Valley, CA: National Water Research Institute.

Parkin, R. T., M. A. Embrey, and P. R. Hunter. 2003. Communicating water–related health risks: lessons learned and emerging issues. Journal of the American Water Works Association 95(7):58–66.

Parks, N. 2003. Fresh approaches to freshwater research. BioScience 53(3):218.

Patten, D. T., and L. E. Stevens. 2001. A managed flood on the Colorado River: background, objectives, design and implementation. Ecological Applications 11(3):635–643.

Pban, K., V. Dushenkov, H. Motto, and I. Raskin. 1995. Phytoextraction—the use of plants to remove heavy metals from soils. Environmental Science and Technology 29(5):1232–1238.

Peterson, G. 2000. Political ecology and ecological resilience: an integration of human and ecological dynamics. Ecological Economics 35:323–336.

Poff, N. L., J. D. Allan, M. B. Bain, J. R. Karr, K. L. Prestegaard, B. D. Richter, R. E. Sparks, and J. C. Stromberg . 1997. The natural flow regime. BioScience 47(11):769–784.

Porter, S. D., M. A. Harris, and S. J. Kalkhoff. 2001. Influence of Natural Factors on the Quality of Midwestern Streams and Rivers. Water–Resources Investigations Report 00–4288. U.S. Geological Survey.

Postel, S. 1999. Pillar of Sand. New York: W.W. Norton.

Powell, K. 2002. Open the floodgates! Nature 420:356–358.

Pulwarty, R. S. 2003. Climate and water in the West: science, information and decision–making. Water Resources Update 124:4–12.

Raupach, M. R. 1999. Group report: how is the atmospheric coupling of land surfaces affected by topography, complexity in landscape patterning and the vegetation mosaic? Pp. 177–196 *In* Integrating Hydrology, Ecosystem Dynamics and Biogeochemistry in Complex Landscapes. J. J. D. Tenhunen and P. Kabat (eds.). Chicester, UK: John Wiley and Sons, Ltd.

Richter, B. D., J. V. Baumgartner, R. Wigington, J. David, and D. P. Braun. 1997. How much water does a river need? Freshwater Biology 37:231–249.

Risbey, J. S., and P. H. Stone. 1996. A case study of the adequacy of GCM simulations for input to regional climate change assessments. J. Climate 9:1441–1467.

Saulnier, G. J. 2002. Use of One-on Analysis to Evaluate Total System Performance. ANL–WIS–PA–000004 Rev. 00 ICN 00. Las Vegas, NV: Bechtel SAIC Company.

Shukla, J., L. Marx, D. Paolino, D. Straus, J. Anderson, J. Ploshay, D. Baumhefner, J. Tribbia, C. Brankovic, T. Palmer, Y. Chang, S. Schubert, M. Suarez, and E. Kalnay. 2000. Dynamical seasonal prediction. Bull. Amer. Meteor. Soc. 81(11):2593–2606.

Slovic, P. 2000. The Perception of Risk. London: Earthscan Publications.

Stakhiv, E. Z. 2003. Disintegrated water resources management. Journal of Water Resources Planning and Management 129:151–155.

Stevens, L. E., T. J. Ayers, J. B. Bennett, K. Christensen, M. J. C. Kearsley, V. J. Meretsky, A. M. Phillips, R. A. Parnell, J. Spence, M. K. Sogge, A. E. Springer, and D. L. Wegner. 2001. Planned flooding and Colorado River riparian tradeoffs downstream from the Glen Canyon Dam, Arizona. Ecological Applications 11(3):701–710.

Stokes, D. E. 1997. Pasteur's Quadrant: Basic Science and Technological Innovation. Washington, DC: Brookings Institution Press.

Strauss, D. 1993. The midlatitude development of regional errors in a global GCM. Journal of the Atmospheric Sciences 50(16):2785–2799.

Terry, N., and G. Banuelos (eds.). 2000. Phytoremediation of Contaminated Soil and Water. Boca Raton, FL: Lewis Publishers.

Turner, R. E., A. M. Redmond, and J. B. Zedler. 2001. Count it by acre or function—mitigation adds up to net loss of wetlands. National Wetlands Newsletter 2(6):5–6,15–16.

U.S. Geological Survey (USGS). 1999. The Quality of Our Nation's Waters—Nutrients and Pesticides. U.S. Geological Survey Circular 1125. 82 p.

Valentini, R., D. D. Baldocchi, and J. D. Tenhunen. 1999. Ecological controls on land–surface atmospheric interactions. Pp. 117–145 *In* Integrating Hydrology, Ecosystem Dynamics and Biogeochemistry in Complex Landscapes. J. D. Tenhunen and P. Kabat (eds.). Chicester, UK: John Wiley and Sons, Ltd.

Wang, X. L., and F. W. Zwiers. 1999. Interannual variability of precipitation in an ensemble of AMIP climate simulations conducted with the CCC GCM2. Journal of Climate 12:1322–1335.

Webb, R. H., J. C. Schmidt, G. R. Marzolf, and R. A. Valdez (eds.). 1999. The controlled flood in Grand Canyon. Geophysical Monograph 110. Washington, DC: American Geophysical Union.

Wieringa, M. J., and A. G. Morton. 1996. Hydropower, adaptive management and biodiversity. Environmental Management 20:831–840.

Wilchfort, G., and J. R. Lund. 1997. Shortage management modeling for urban water supply systems, Journal of Water Resources Planning and Management-ASCE 123(4):250–258.

Williams, B. L., and Y. Florez. 2002. Do Mexican-Americans perceive environmental issues differently than Caucasians: a study of cross-ethnic variation in perceptions related to water in Tucson. Environmental Health Perspectives 110(S2):303–310.

4

Status and Evaluation of Water Resources Research in the United States

Establishing a baseline of current research is vital to the task of evaluating whether and how new priorities in water resources research are being addressed. Research is a cumulative enterprise. By necessity, most new research directions will build on existing research infrastructure; other research directions may be established through new research consortia, laboratories, and field sites. Whatever the case, budget initiatives will be cast in terms of departures from the status quo. Unfortunately, the categorization and accounting of water resources research is surprisingly difficult to do under current budgetary procedures. Agencies are not required to report their research to the Office of Management and Budget (OMB) in standard topical or thematic categories. Further, agencies do not report all their research to OMB. For this reason, the committee gathered budget[1] and other data, in the form of a survey, from both federal agencies and nonfederal organizations that fund water resources research. This chapter presents the resulting data and the committee's analysis of those data, as well as conclusions about the scope of the current investment. The conclusions relate directly to those water resources research priorities expressed in Chapter 3 and in NRC (2001) as being paramount to confronting water problems that will emerge in the next 10–15 years.

SURVEY OF WATER RESOURCES RESEARCH

A necessary part of this study involved collecting budget information from federal agencies and significant nonfederal organizations regarding their recent

[1]Budget in this chapter refers to actual expenditures, unless otherwise noted.

expenditures on water resources research. Several methods could potentially be utilized to gather and evaluate such budget information. Ultimately, the committee decided to rely on a format used for over ten years in the late 1960s and early 1970s.

Beginning in 1965, the Committee on Water Resources Research (COWRR) of the Federal Council for Science and Technology (FCST), administered out of what was then the President's Office of Science and Technology, began a yearly accounting of all water resources research conducted by the major federal agencies.[2] Budget information, supplied by liaisons from relevant federal agencies, was compiled into ten major categories[3] and up to 60 comprehensive subcategories of water resources research. The accounting occurred annually from 1965 to 1975 (except for 1971). The primary goal of COWRR was to facilitate coordination of the various federal research efforts, because it was recognized at that time that water resources research was spread widely throughout the federal enterprise (as it is today). It was also a goal of COWRR to ensure that there was no unnecessary duplication of research efforts, that research was appropriately responsive to current water problems, and that federal resources were available to help solve these problems (COWRR, 1973 and 1974). Nonfederal organizations were not included in the reports.

To compare the current budget information with expenditures on water resources research between 1965 and 1975, the committee adopted the FCCSET model of creating a survey for federal agency liaisons to respond to. The present survey includes most of the same categories and subcategories of water resources research as before, and it encompasses the same waterbodies: fresh, estuarine, and coastal. In January 2003, the survey was submitted to all of the federal agencies that either perform or fund water resources research and to several nonfederal organizations that had annual expenditures of at least $3 million during one of the fiscal years covered by the survey.

The survey consisted of five questions related to water resources research (see Box 4-1). As part of question 1, the liaisons were asked to report total expenditures on research in fiscal years 1999, 2000, and 2001, in order to allow a comparison to the FCCSET survey data of the past. The remaining four questions were posed to help give the committee a better understanding of current and projected future activities of the agencies, and to obtain a qualitative understanding of how research performance is measured. Unlike the data submitted in response to question #1, the answers to the latter questions in the survey are not evaluated in this report in a *quantitative* fashion.

Responses to the survey were submitted in written form and orally at the third meeting of the NRC committee, held April 29–May 1, 2003, in Washington,

---

[2]In 1976, COWRR came under the aegis of the Federal Coordinating Council of Science, Engineering, and Technology (FCCSET) of the Office of Science and Technology Policy.

[3]There were only nine major FCCSET categories from 1965 to 1970. A tenth (Scientific and Technical Information) was added in 1972.

BOX 4-1
Survey of Federal Liaisons on Water Resources Research

1a. Please provide budget information for the 11 FCCSET categories for FY1999, FY2000, and FY2001 (as total expenditures, not appropriated funding). A detailed description of each category is attached.

b. Please provide an accompanying short (2–3 pages at most) narrative, saying how your programs that encompass water resources research fall into the different FCCSET categories.

c. What percentage of these budget numbers were reported to OMB as R&D? (We recognize that there are differences between the OMB definition of R&D and the 11 FCCSET categories.)

d. Does your agency conduct research that does not fall into one of the FCCSET categories, but that is considered (by the agency) to be water resources research? Please describe.

2. In no more than 2 pages, provide a summary of your agency's current strategic plan that governs water resources research. Please include data collection activities.

3. What is being done to coordinate water resources research (1) within your agency, (2) with other agencies, or (3) with external partners (such as the states)?

4. Do you measure progress (i.e., the impact of) in your agency's water resources research activities? If so, how? (For example, by counting the number of publications, some other metric, etc.)

5. Irrespective of your agency's mission, what do you think the nation's water research priorities ought to be?

- Identify the major water issues that will confront the nation in the next 5 to 10 years and research topics that would be helpful in addressing those major issues.
- Point out gaps in current data collection system to address these priorities.

D.C. At that meeting, questions were asked of the liaisons in addition to those listed in Box 4-1 that speak to the different ways that research is conceptualized and conducted within the federal enterprise. These included questions about (1) how the budget information was gathered and the liaison's confidence in its accuracy, (2) whether the water resources research included in the survey response was conducted internally or externally, and (3) the typical time frame for water resources research within an agency. (Definitions of research relevant to these questions are provided in Box 3-1.) Revised survey responses submitted by the liaisons in June and July 2003 reflected corrections and responses to specific requests from the committee.

The survey requested budget information in 11 major categories (and 71 subcategories) of water resources research. All categories, which are described in detail in Appendix A, closely correspond to categories used in the old FCCSET reports. Nonetheless, some minor changes were made to the old FCCSET categories in order to capture lines of research that were not recognized during the 1960s. Most importantly, new subcategories were added in the areas of global water cycle problems, effects of waterborne pollution on human health, risk perception and communication, other poorly represented social sciences, infrastructure repair and rehabilitation, restoration engineering, and facility protection/national security. One of the old FCCSET subcategories (VI-C on the ecological impact of water development) was removed from Category VI and was expanded into a new major category (XI) that includes four subcategories on ecosystem and habitat conservation, aquatic ecosystem assessment, effects of climate change, and biogeochemical cycles. This was done in recognition of the increased attention being paid to the water needs of aquatic ecosystems over the last 25 years and a corresponding surge in research in this field. The modified FCCSET categories thus comprehensively describe all areas of research in water resources. It should be noted that the act of data collection, although of paramount importance to water management, is not captured by any of the modified FCCSET categories (Category VII covers research that informs data collection, not data collection itself). This omission on the part of COWRR was intentional, allowing research activities and their budgets to be evaluated independently of monitoring activities. The current survey abides by this separation; that is, the agency liaisons made sure that none of the budget information presented includes pure data collection (e.g., stream gaging, satellite operation, etc.) Nonetheless, given the importance of data collection activities to the water resources research enterprise, Chapter 5 notes recent trends in funding for such activities.

There are obviously limitations inherent in conducting a survey of this nature and in the corresponding results. First and foremost is that the information represents to some degree the best professional judgment of those liaisons that responded. In almost all cases, federal agency programs in water resources research are not organized along the modified FCCSET categories. Undoubtedly, there were cases where a program logically fell into more than one category. In such cases, the liaisons were asked to give their best judgment of the most relevant category. In addition, variable sources of information were used by the liaisons (in terms of personnel and databases consulted), and the liaisons may have interpreted the survey differently from one another. These factors are reflected by a certain degree of error in the individual budget numbers submitted by the liaisons. However, after questioning the liaisons about their confidence in the submitted information, the committee feels that the magnitude of this error is small when compared to the broad trends that are discerned by the analysis below. Furthermore, the trend analysis is accompanied by a quantitative assessment of uncertainty, which was taken into account during the committee's evaluation of the data.

Second, the possibility exists that the committee did not capture all of the relevant federal and nonfederal organizations involved in water resources research, either because these organizations were not approached by the committee or because they chose not to participate. With respect to the federal agencies, the committee is confident that all of the major agencies funding or conducting water resources research within the United States were contacted and that the submitted survey responses represent the vast majority of the federal investment in water resources research. There is less certainty about the nonfederal organizations. The major not-for-profit organizations involved in water resources research were contacted, as well as the largest (in terms of funding) of the state Water Resources Research Institutes (in order to reflect state funds spent on relevant research). Nonetheless, it is recognized later in this chapter that the accounting of significant nonfederal organizations' funding of water resources research may be an underestimate, both in terms of total dollars and represented subcategories.

A related issue for those federal agencies that responded to the survey is that not all of their relevant research funds were reported, especially where certain programs are not characterized as research in their congressional authorization. For example, the Department of Energy's (DOE) site characterization work in the Yucca Mountain Program (see Chapter 3) is at the cutting edge of hydrogeology, but it is not classified as research for budgetary purposes. [In contrast, the U.S. Geological Survey (USGS) does classify its Yucca Mountain work as research.] Because the agencies would be reluctant to report these types of expenditures, it was not possible for the committee to assess their magnitude or importance. This is also a concern for agencies that conduct extensive place-based studies, most of which are managed separately from the general water resources research program, are not reported to OMB as research, and thus are difficult to account for. Examples include the Florida Everglades restoration—jointly run by the U.S. Army Corps of Engineers (Corps), the Department of the Interior, and the South Florida Water Management District—and CALFED, which is a San Francisco Bay Delta restoration program involving multiple federal and state agencies. For those federal agencies that were identified at the April 2003 committee meeting as funding substantial place-based research, the committee requested that their two largest projects be included in their final response to the survey. These inclusions are reflected in revised survey responses from the Corps and the U.S. Bureau of Reclamation. Nonetheless, not all place-based research from these two agencies could be captured, and no place-based information was collected from other agencies. The U.S. Environmental Protection Agency (EPA) is the other major federal agency thought to have an investment in place-based research. This may also lead to an underestimate in the reported water resources research funding.

Third, the survey covers only fiscal years 1999, 2000, and 2001. Three years of data were felt to be of sufficient quantity to allow the committee to assess the nation's investment without creating a burdensome task for the liaisons. In addition, vastly differing economic climates prevailed during these years, which may be reflected in the survey responses and may thus enable the committee to observe short-term variability in research expenditures. Clearly, however, these three years of data represent only a snapshot in time. Thus, although there is a trends analysis in this chapter, no assumptions should be made regarding funds spent between 1975 and 1999.[4] The current request for information did not cover FY2002 or FY2003 because it was felt that at the time the survey was submitted, the agencies would not be able to provide accurate estimates of total expenditures for those years. Thus, events subsequent to FY2001 that may have impacted research spending (e.g., increased attention to national security) are not reflected in the survey.

Finally, the varying scope of the modified FCCSET categories must be acknowledged. In an attempt to keep the number of subcategories reasonable, some of them broadly lump together what may be, in academic circles, disparate research issues. For example, there is only a single subcategory (VI-H) to capture all water resources research conducted in areas of sociology, anthropology, geography, political science, and psychology. Other subcategories are much more narrowly focused. This diversity, to a certain extent, reflects the fact that some subcategories have a stronger historical linkage to water resources research per se. In general, in those areas where the majority of funds are being spent, the committee tried to maintain or create a larger number of subcategories so that specific trends in funding could be discerned.

For the purposes of the discussion below, the budget numbers from all years were converted into FY2000 constant dollars prior to graph preparation and data analysis.

OLD FCCSET DATA

From 1965 until 1975, data on water resources research funds were collected from the following federal agencies: U.S. Departments of Agriculture, Commerce,

[4]There are estimates for total spending on water resources research between 1979 and 1987 from the 1980 report "U.S. National Water Resources Research Development, Demonstration, and Technology Transfer Program 1982–1987" summarized in NRC (1981). However, these estimates are not included in this report because their accuracy could not be verified.

Defense, the Interior, and Transportation; EPA (from 1973 on); the National Science Foundation (NSF); the National Aeronautics and Space Administration (NASA) (from 1966 on); and other smaller agencies such as the Tennessee Valley Authority, Housing and Urban Development, the Atomic Energy Commission, and the Smithsonian Institute. The budget data are presented in a series of annual reports from COWRR and are summarized in COWRR (1973 & 1974). FCCSET data show a steady increase in funding for water resources research between 1964 and 1967, a leveling off from 1967 to 1973, and a slight decrease from 1973 to 1975 (see Figure 4-1). A more in-depth examination reveals that the vast majority of these funds were spent in a few FCCSET categories, and these disparities increased during the examined period. Thus, for example, in 1965, Categories II (water cycle), III (water supply augmentation and conservation), and V (water quality management and protection) constituted over 60 percent of all water resources research, while in 1975, these same categories comprised 76 percent of the total. As shown in Figure 4-2, this increase is attributable to a large increase in spending on water quality management and protection (Category V).

The only FCCSET category that showed positive growth during this ten-year period was V (water quality management and protection), and even this category began to decline after 1973. Most of the other major categories showed relatively stagnant funding during the period, including II (water cycle), IV (water quantity management and control), VII (resources data), X (scientific and technical infor-

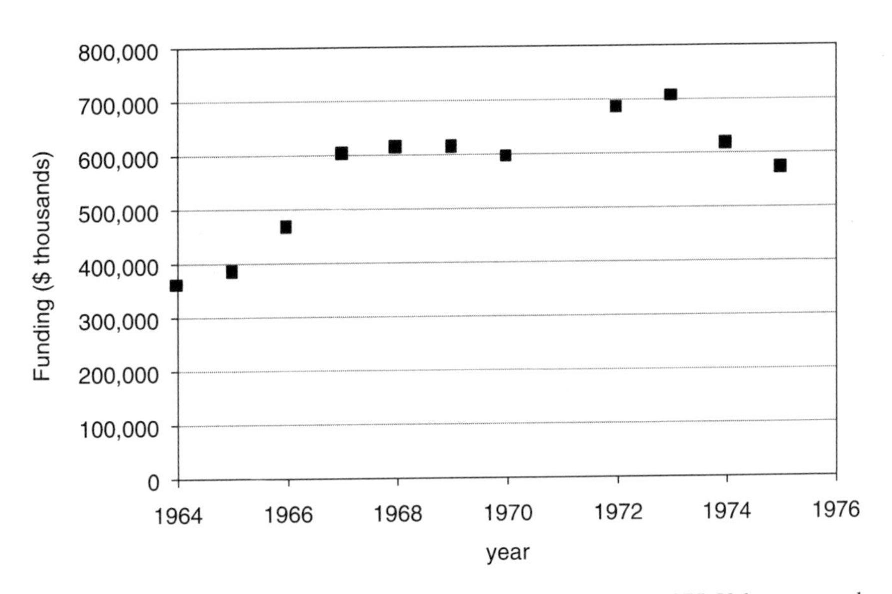

FIGURE 4-1 Total expenditures on water resources research, 1964–1975. Values reported are constant FY2000 dollars. SOURCES: COWRR (1973 & 1974).

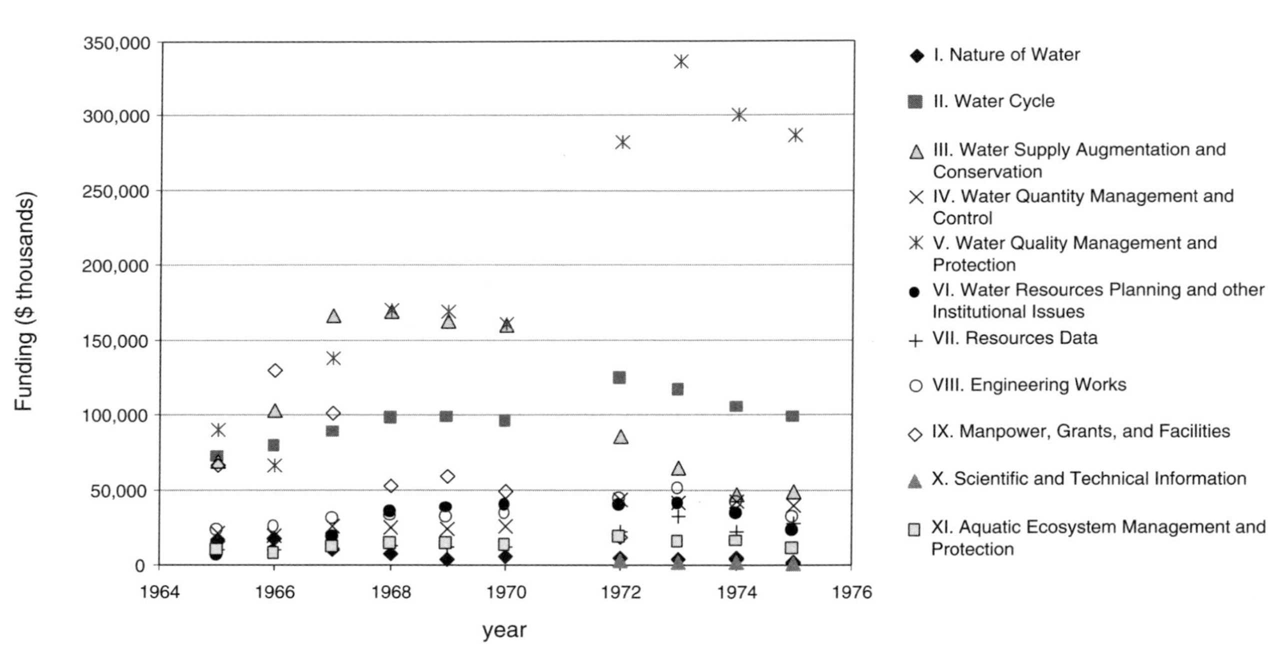

FIGURE 4-2 Funding in all major FCCSET categories, 1965–1975. Values reported are constant FY2000 dollars. SOURCES: COWRR (1973 and 1974).

mation), and XI (aquatic ecosystem management and protection). Consistently negative trends in funding were observed for Categories I (the nature of water) and IX (manpower, grants, and facilities). For Categories III (water supply augmentation and conservation), VI (water resources planning), and VIII (engineering works), an initial increase observed in the late 1960s was followed by a substantial decrease in the early 1970s, in the case of Category III to levels below the 1965 level (primarily because of a drop in funding for desalination research). Trends for the "smaller" major categories, which are difficult to discern in Figure 4-2, are presented in Figure 4-3.

The reasons for the observed trends likely include an initial interest on the part of Congress and various administrations to increase research spending in the late 1950s and early 1960s, followed by a retraction in the wake of better understanding environmental processes and the resulting competition between environmental water needs and economic growth. As discussed in Chapter 2, the early 1970s saw the federal government transform from an ardent supporter of water resources projects to the primary regulator of industries responsible for declines in water quality. This may also account for the disproportionate support for water quality research (Category V) compared to other areas of research. That is, greater investment in Category V was seen as essential to meeting various water quality standards in the nation's lakes and rivers, as mandated by the newly minted Clean Water and Safe Drinking Water Acts. Furthermore, in many states, impairment in water quality loomed as a more important constraint on the development of water resources than the issue of supply. In addition, throughout the 1970s, media reports focused on water quality issues, giving them the political prominence that has helped to drive the distribution of research funding shown in Figure 4-2. The topically skewed nature of water resources research in the middle 1970s has been noted in other studies, in particular a FCCSET report that recommended reducing the relative proportion of funding going to Category V, while also calling for overall increases in the total water resources research budget (COWRR, 1977).

WATER RESOURCES RESEARCH FROM 1965 TO 2001

To observe trends in water resources research funding, the FCCSET accounting of 1965–1975 was repeated by requesting budget information from 19 federal agencies known to support water resources research. Table 4-1 lists the federal agencies queried during the first survey period and during this study. A similar request was made of several nonfederal organizations, of which the following were deemed to be making significant contributions to water resources research over the period in question (FY1999–FY2001): the American Water Works Association Research Foundation (AWWARF), the Water Environment Research Foundation (WERF), the Nature Conservancy (TNC), and the four largest Water Resources Research Institutes (Nevada, Pennsylvania, Texas, and Utah). For both

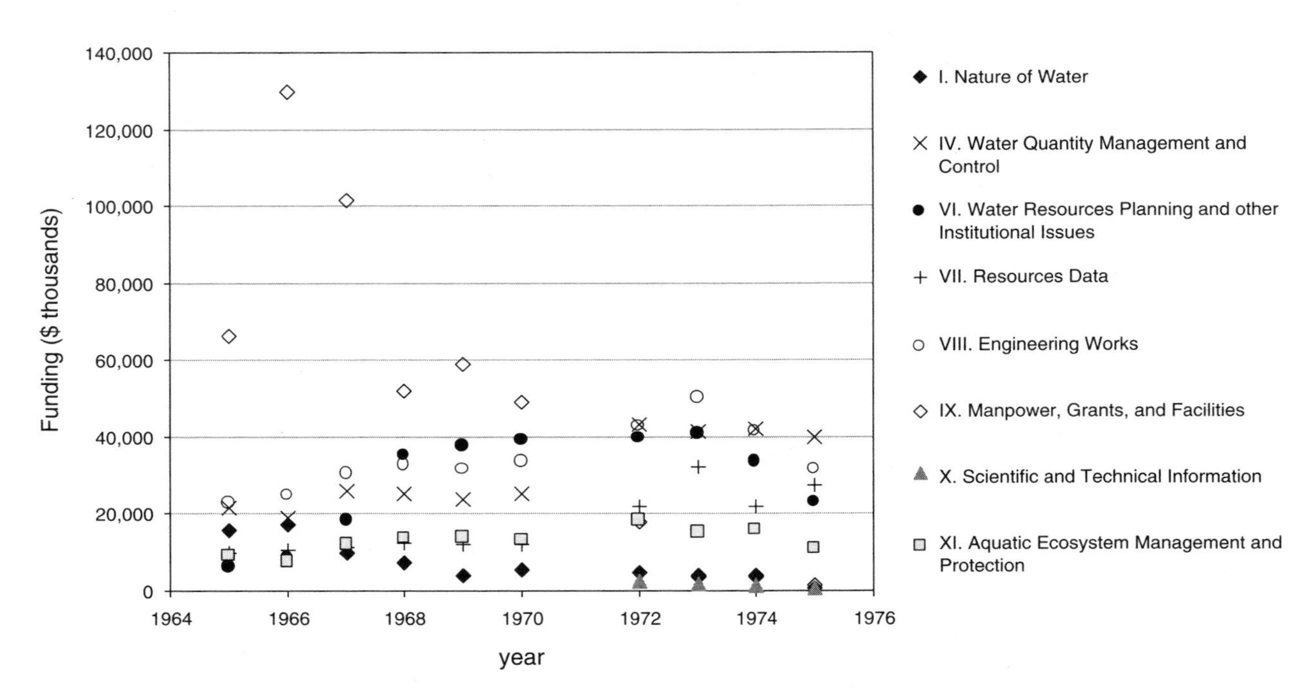

FIGURE 4-3 Funding in the "smaller" major FCCSET categories, 1965–1975. Values reported are constant FY2000 dollars. SOURCES: COWRR (1973 & 1974).

TABLE 4-1 Federal Agency Participation in Surveys on Water Resources Research Funding

Agency	Initial FCCSET period (1965–1975)	Current Survey (FY1999–FY2001)
Agriculture		
ARS	Yes	Yes
CSREES	Yes	Yes
ERS	Yes	Yes
FS	Yes	Yes
Commerce		
NOAA (many programs)	Yes	Yes
Defense		
Corps	Yes	Yes
ONR	No	Yes
SERDP/ESTCP	No	Yes
Energy	No	Yes
Health and Human Services	No	
ATSDR		Yes
NCI		Yes
NIEHS		Yes
Interior		
USGS	Yes	Yes
USBR	Yes	Yes
FWS	Yes	No
OWRR	Yes	No longer in existence
Transportation	Yes (1966–1971)	No
FHA	Yes (1973–1975)	
Coast Guard	Yes (1973–1975)	
EPA	Yes (1973–1975)	Yes
NASA	Yes	Yes
NSF	Yes	Yes
AEC[a]	Yes	No longer in existence
TVA	Yes	No
Smithsonian	Yes (1968–1975)	No
HUD	Yes (1967–1975)	No

[a]The functions of the Atomic Energy Commission were subsumed by the Nuclear Regulatory Commission via the Energy Reorganization Act of 1974 and by DOE.

Note: The National Park Service, U.S. Fish and Wildlife Service, and the Tennessee Valley Authority were contacted but chose not to participate in the current survey.

the federal agencies and the nonfederal organizations, the budget information submitted was total expenditures (and not appropriated funds). In addition, all third-party funding was excluded from the budget numbers, as was funding for pure data collection, education, and extension activities. For the federal agencies, almost all of the funds included in the survey are reported to OMB as research and development funds. Survey data from both federal and nonfederal agencies are summarized and presented graphically throughout the chapter, with detailed information available in Appendix B.

Total Federal Agency Support of Water Resources Research

The total federal funding for water resources research from 1964 to 1975 and for 1999, 2000, and 2001 is shown in Figure 4-4. Annual expenditures on water resources research have remained static near the $700 million mark since 1973, after having doubled between 1964 and 1973. A quantitative analysis was conducted to discern whether there is a significant difference in total funding between the average 1973–1975 levels and the average 1999–2001 levels. In order to do this analysis (which is explained in detail in Appendix C), it was assumed that the annual data contain measurement errors that are independent from year to year, that the distribution of errors in averages of annual values can be well approximated by a normal distribution, that the standard deviation of the errors in averages of annual values ranges in all cases from 25 percent to 50 percent of the

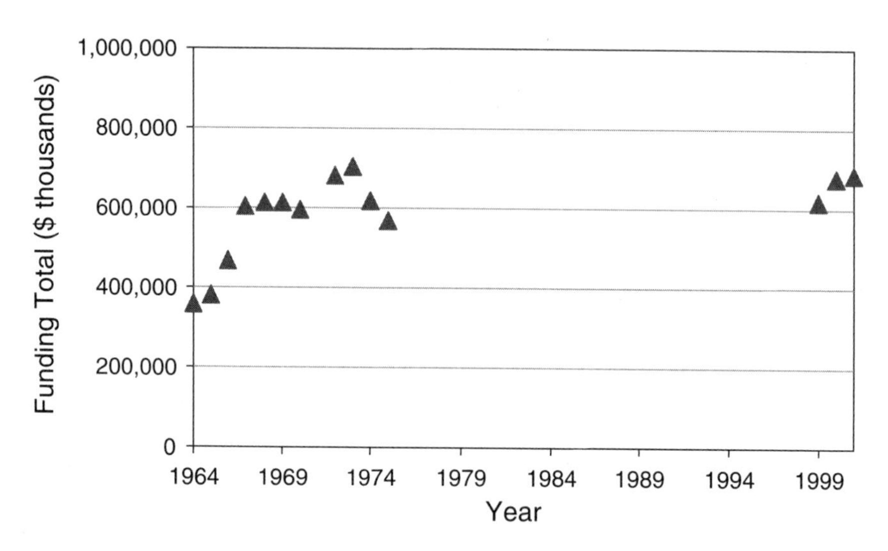

FIGURE 4-4 Total expenditures on water resources research by federal agencies, 1964–2001. Values reported are constant FY2000 dollars.

average, and that there are no significant systematic biases in the annual funding data. As shown in Table 4-2, there is small likelihood that the averages for these two time periods are different under the conditions of uncertainty stated above. This supports the statement that funding levels for water resources research have not changed significantly since the early 1970s.

Figure 4-5 shows how the $700 million was distributed among the federal agencies in FY2000. Notably, no agency contributed more than 25 percent of the total funding, with five agencies (USDA, USGS, NSF, DoD, and EPA) accounting for nearly 88 percent of the total funding. The smaller five agencies (DHHS, USBR, NOAA, DOE, and NASA) together contributed 12.3 percent of the reported total. This speaks to the broad impact of, and interest in, water-related issues across the federal government. Funding trends from 1965 to 2001 for the 11 major categories and their subcategories are shown in Figures 4-6 and 4-7,

TABLE 4-2 Likelihood of a Significant Increase or Decrease in Funding from the 1973–1975 Time Period to the 1999–2001 Time Period (see Appendix C for methods)

FCCSET Category	Likelihood at the 25 percent uncertainty level (%)	Likelihood at the 50 percent uncertainty level (%)	Increase or Decrease*
I. Nature of Water	99.9	92.6	Large increase
II. Water Cycle	79.3	66.1	Increase
III. Water Supply Augmentation	0.2	7.5	Large decrease
IV. Water Quantity	59.3	55.1	No change
V. Water Quality	12.0	28.0	Large decrease
VI. Water Resources Planning	0.2	7.5	Large decrease
VII. Resources Data	0.5	9.8	Large decrease
VIII. Engineering Works	66.6	57.9	Slight Increase
IX. Manpower, Grants, Facilities	100.0	95.0	Large increase
X. Scientific and Technical Info	45.2	46.8	Slight decrease
XI. Aquatic Ecosystems	100.0	96.7	Large increase
Total Water Resources Research	55.1	52.5	No change
Total Water Resources Research minus Category XI	28.5	38.9	Decrease

*Values above 50 indicate a significant increase from the mid 1970s to the late 1990s. Values less than 50 indicate a significant decrease. Values around 50 percent indicate no significant increase or decrease.

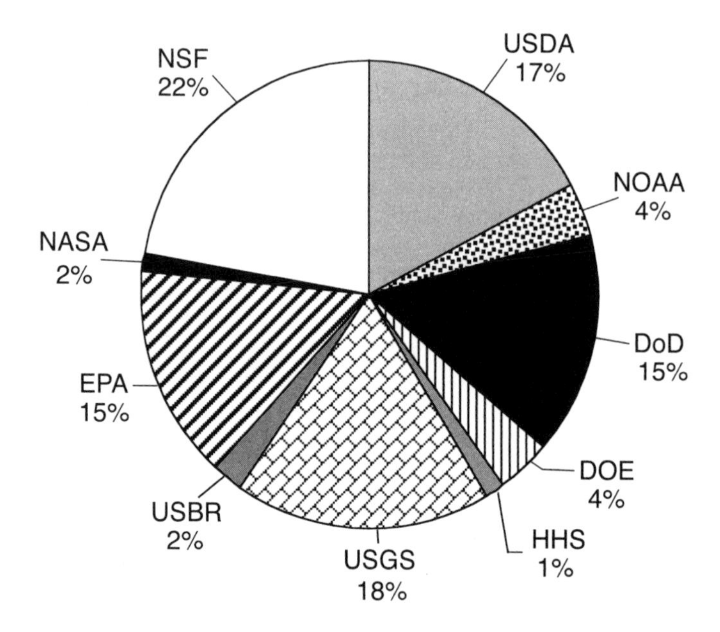

FIGURE 4-5 Federal agency contributions as a percentage of the total funding for water resources research in 2000.

respectively. Note that all three years of data (1999, 2000, and 2001) were used in the trends analysis to follow. However, for the purposes of presentation, the pie charts in this chapter show only data from 2000 (which is generally representative of data from 1999 and 2001).

Several conclusions can be drawn from these graphs. With respect to the trends over time for the individual major categories, most funding levels have remained stable or have declined since the mid 1970s—a conclusion supported by the comprehensive uncertainty analysis presented in Table 4-2 and in Appendix C. Category V (water quality), in particular, declined from 1975 to 2000 both in real terms (from $286 million to $192 million) and as a percentage of total funding (from 50 percent to 28 percent). For this category, it can be stated with high confidence that the mid 1970s funding is higher than the late 1990s funding. Even more dramatic declines are observed for Category III (water supply augmentation), which declined from a high of $64 million in 1973 to $14 million in 2000; Category VI (water resources planning and institutional issues), which declined from a high of $41 million in 1973 to $9.8 million in 2000; and Category VII (resources data), which declined from a high of $32 million in 1973 to $8.7 million in 2000. In these three cases, there is very high likelihood that the mid 1970s

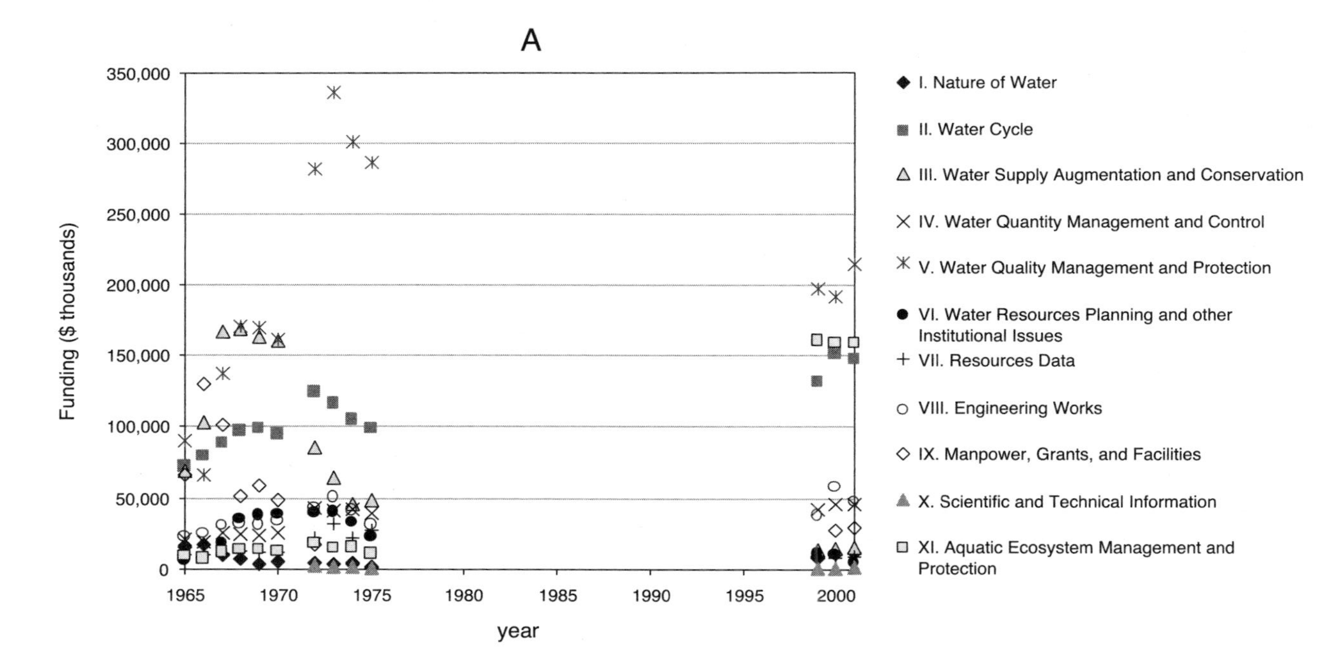

FIGURE 4-6 Federal agency funding in the major modified FCCSET categories, 1965–2001. (A) shows the trends on an arithmetic scale. Values reported are constant FY2000 dollars.

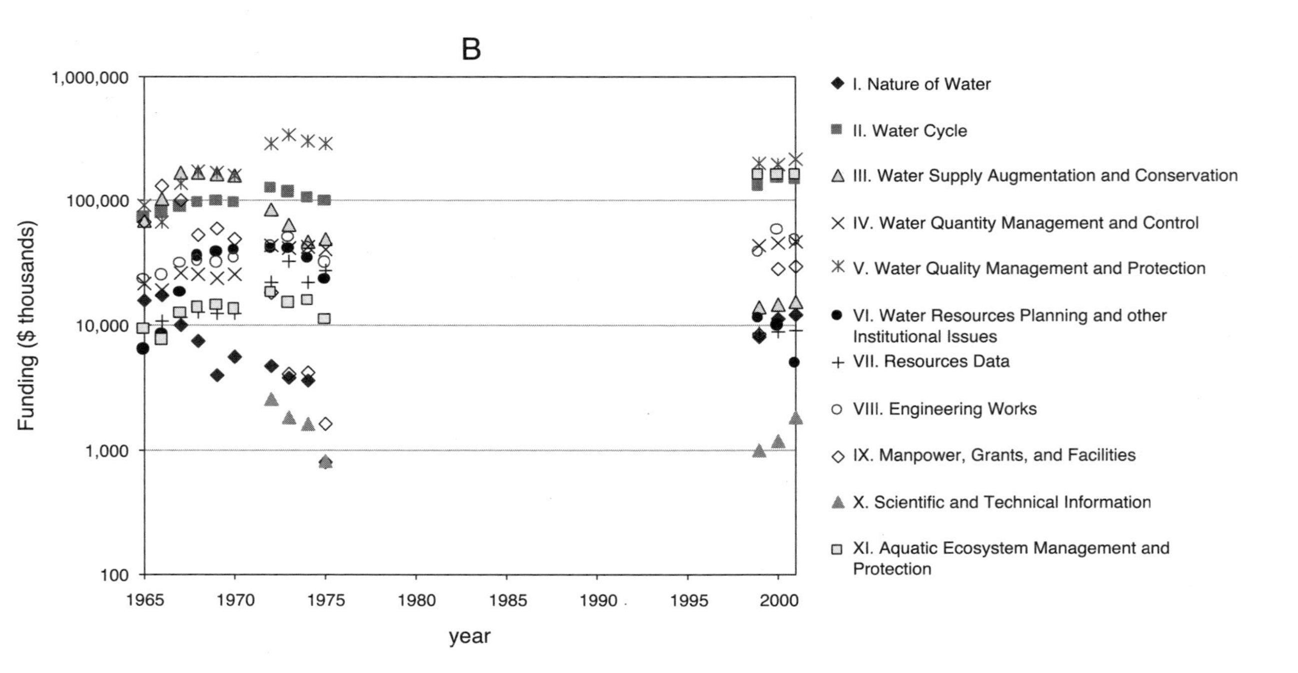

FIGURE 4-6 Federal agency funding in the major modified FCCSET categories, 1965–2001. (B) uses a logarithmic scale. Values reported are constant FY2000 dollars.

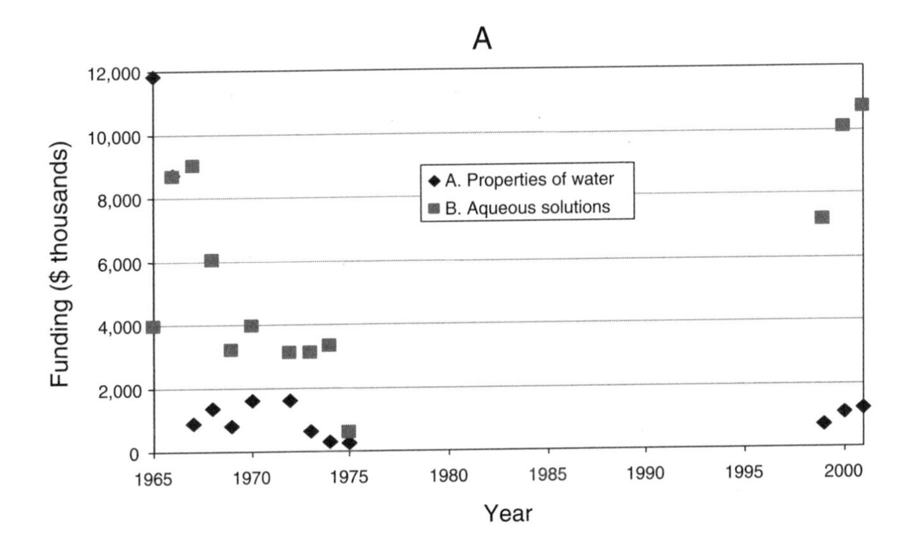

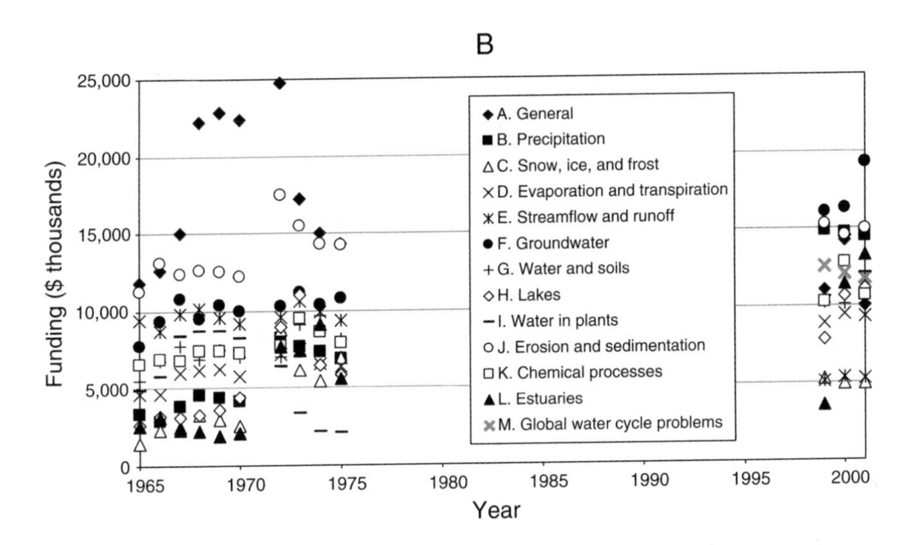

FIGURE 4-7 (A) and (B) Federal agency funding in FCCSET subcategories of Categories I (nature of water) and II (water cycle), 1965–2001. Subcategory II-M is new to the recent survey. Values reported are constant FY2000 dollars.

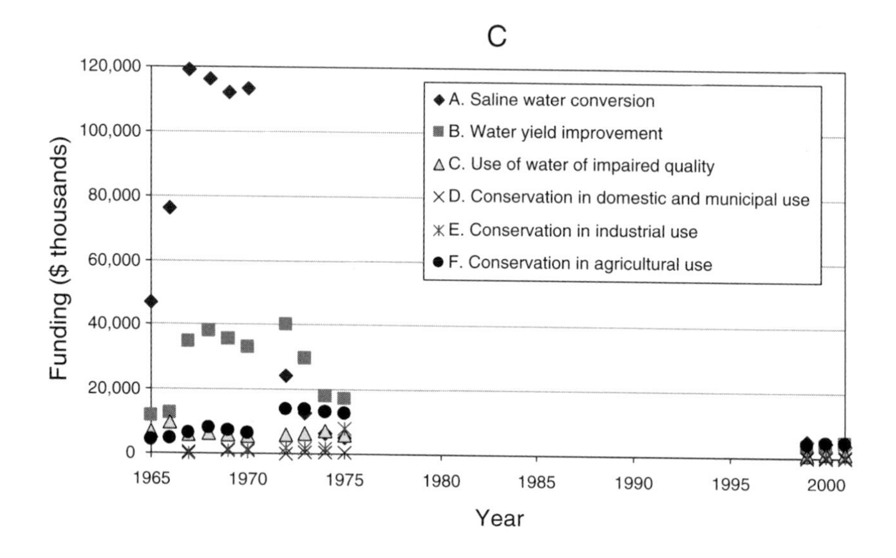

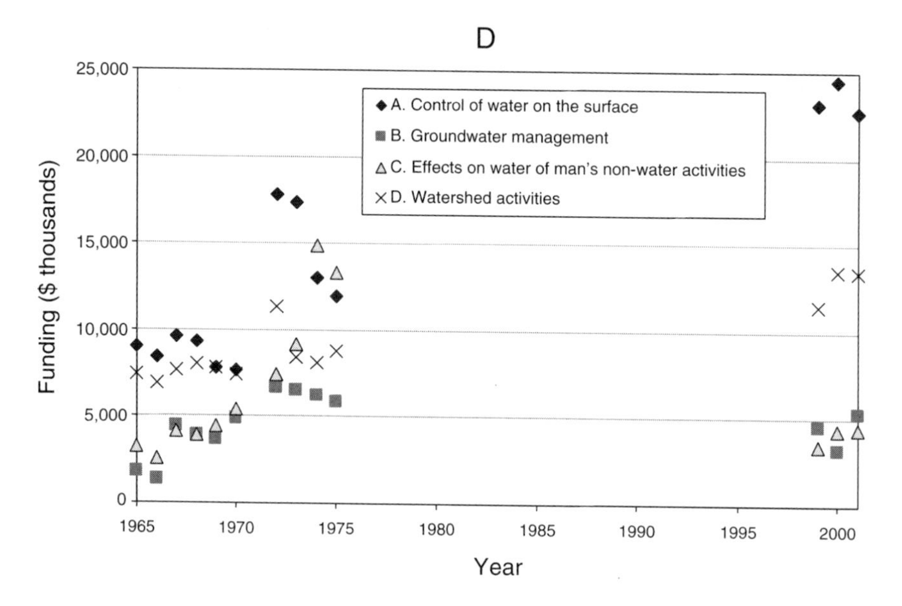

FIGURE 4-7 (C) and (D) Federal agency funding in FCCSET subcategories of Categories III (water supply augmentation and conservation) and IV (water quantity management and control), 1965–2001. Values reported are constant FY2000 dollars.

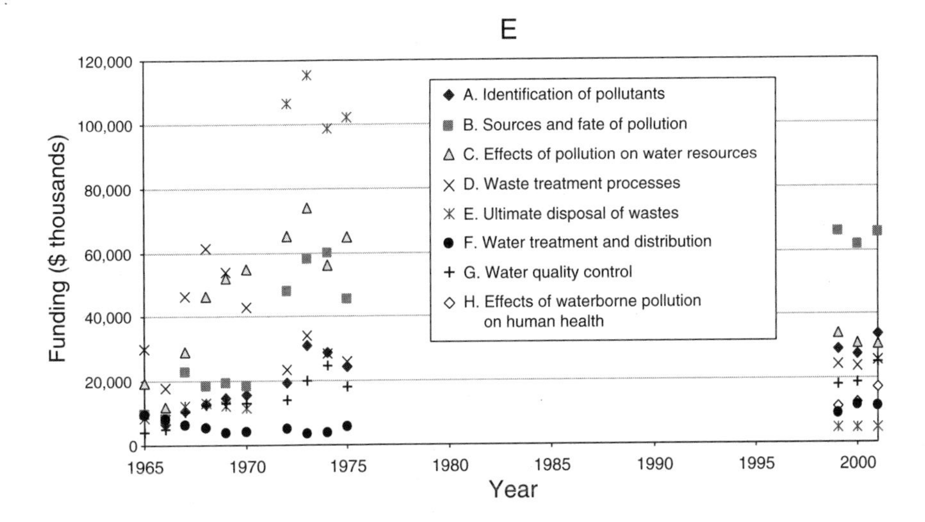

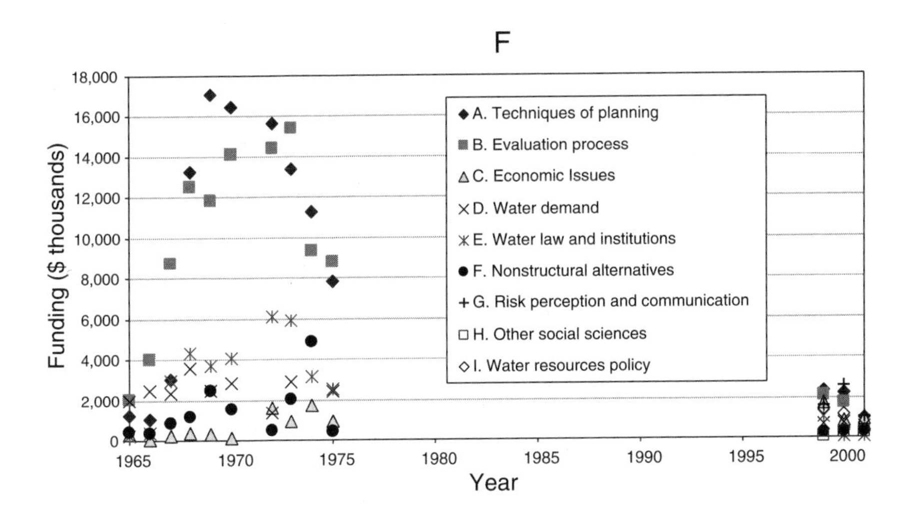

FIGURE 4-7 (E) and (F) Federal agency funding in FCCSET subcategories of Categories V (water quality management and protection) and VI (water resources planning and other institutional issues), 1965–2001. Subcategories V-H, VI-G, VI-H, and VI-I are new to the recent survey. Values reported are constant FY2000 dollars.

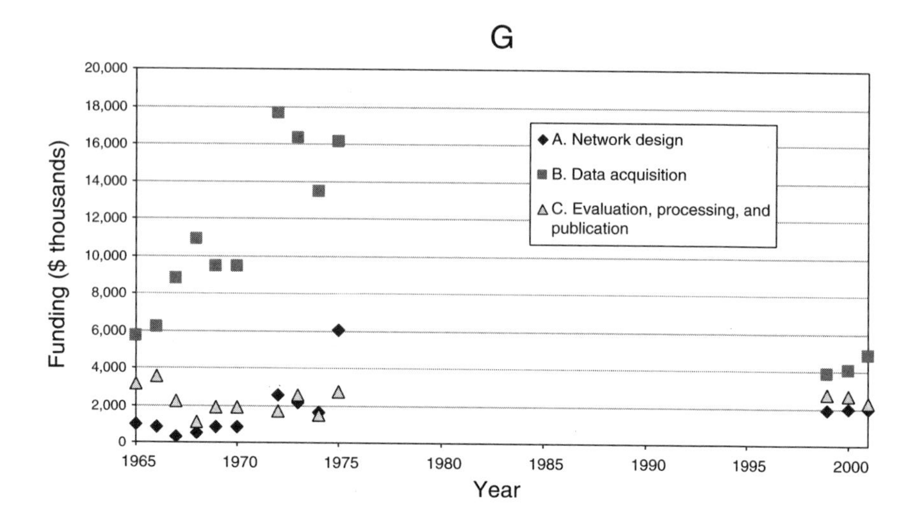

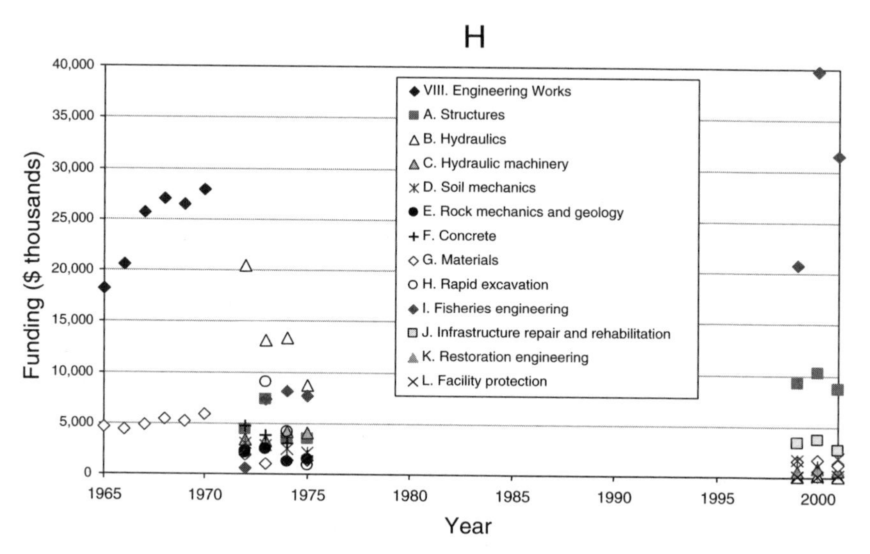

FIGURE 4-7 (G) and (H) Federal agency funding in FCCSET subcategories of Categories VII (resources data) and VIII (engineering works), 1965–2001. Subcategories VIII-B through VIII-I (except VIII-G) were created in 1972 to further define "engineering works." Subcategories VIII-J, -K, and -L are new to the recent survey. Values reported are constant FY2000 dollars.

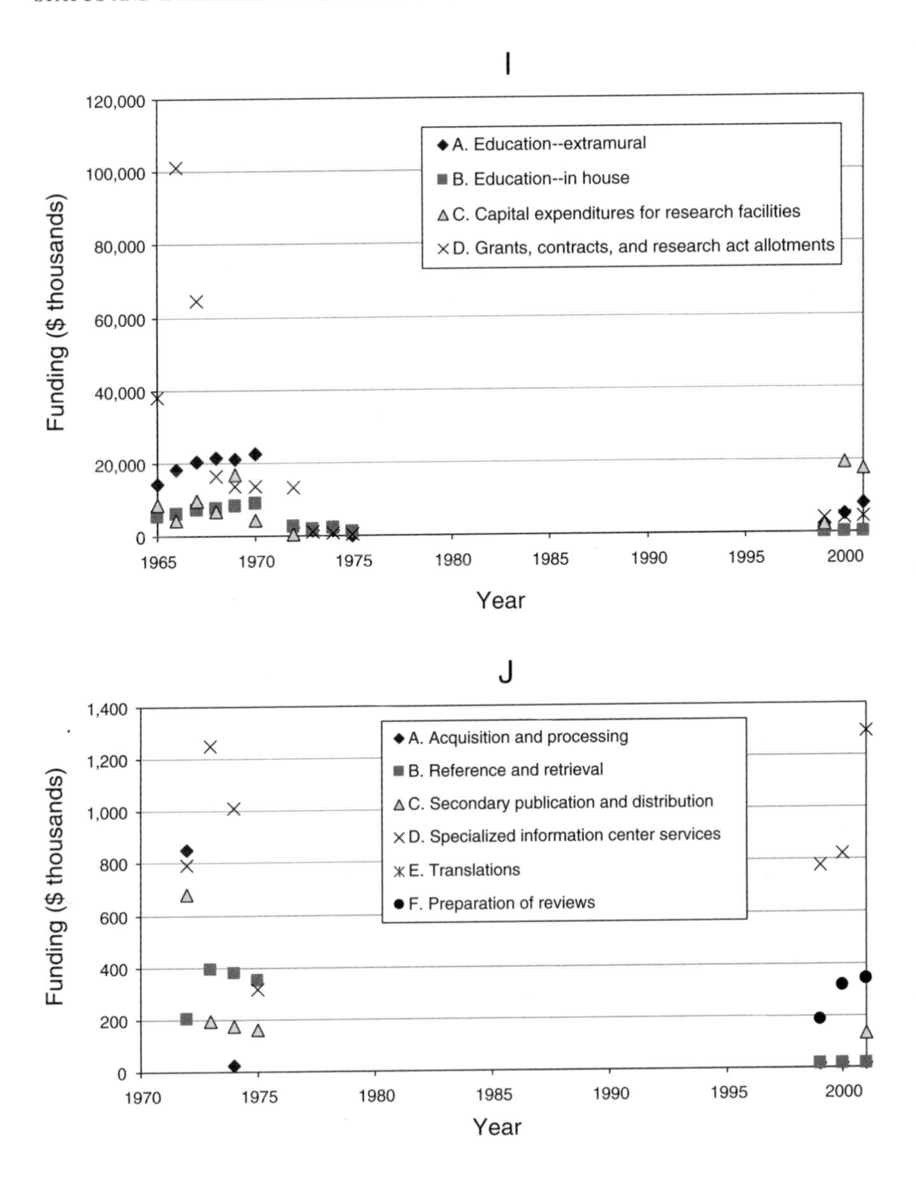

FIGURE 4-7 (I) and (J) Federal agency funding in the FCCSET subcategories of Categories IX (manpower, grants, and facilities) and X (scientific and technical information), 1965–2001. Category X was created in 1972. Values reported are constant FY2000 dollars.

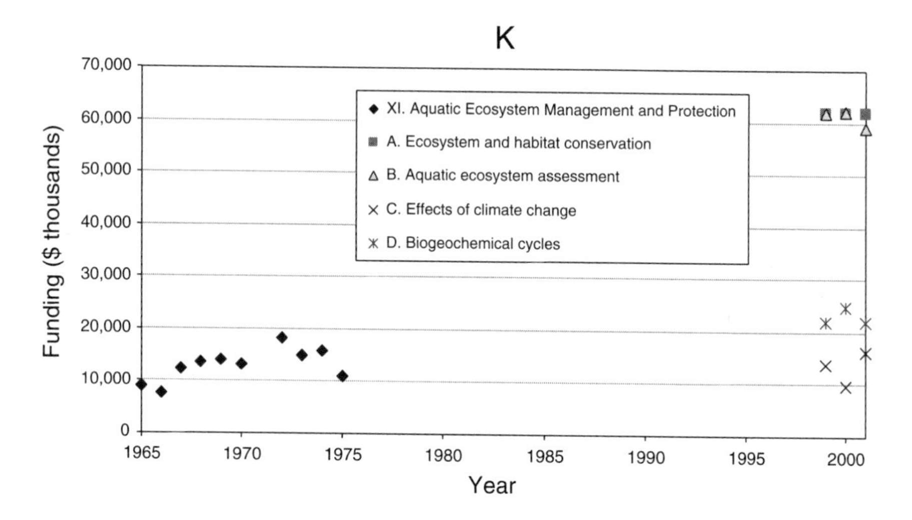

FIGURE 4-7 (K) Federal agency funding in the FCCSET subcategories of Category XI (aquatic ecosystem management and protection), 1965–2001. Note that the four subcategories were created for the recent survey. All relevant research in these four areas was previously recorded in one subcategory (denoted by the black diamond). Values reported are constant FY2000 dollars.

level funding is higher than the late 1990s level, for both cases of uncertainty depicted (that is, where uncertainty is either 25 percent or 50 percent of the mean values). Thus, funding has declined substantially in the late 1990s with respect to the mid 1970s for these four major categories.

Funding in five of the major categories (I—nature of water, IV—water quantity, VIII—engineering works, IX—manpower and grants, X—scientific and technical information) appears from the figures above to be more or less comparable to that recorded 30 years ago. However, when looking quantitatively at the difference between the 1973–1975 and 1999–2001 time periods, some minor trends emerge. In the case of Categories I and IX, there have been significant increases from the mid 1970s to the present. For Category X, one can state with high confidence that funding has decreased since the mid 1970s. And for Categories IV and VIII, there are no significant differences between the funding levels of the two time periods, particularly when it is assumed that uncertainty is equal in value to 50 percent of the mean.

There have been very modest increases in funding for Category II (water cycle) over the entire 30-year period, and the uncertainty analysis in Appendix C supports a high likelihood for an increase in funding from the mid 1970s to the present. However, it is in Category XI where the greatest increases are observed.

Funding for aquatic ecosystem management and protection jumped from $15 million in 1973 to $158 million in 2000 (23 percent of the total water resources research funding that year). This partly reflects the lack of a well-defined category for aquatic ecosystem research during the time of the 1965–1975 FCCSET survey. More important, however, is that concern for aquatic ecosystem management and protection grew enormously following the transformative events that sparked the landmark environmental legislation of the 1960s and 1970s including the Clean Water Act, the Endangered Species Act, and the Wild and Scenic Rivers Act (see Chapter 2). Several decades later, biological diversity and ecosystem processes of lakes, wetlands, and rivers are increasingly at risk, raising concerns for potential degradation of ecosystem goods and services and loss of species. While many of these problems have a long history, most prior research focused on a narrow view of water quality for human use and direct harm to sensitive species. As a consequence of the recent recognition for the need for whole ecosystem research, including studies of long duration and large spatial scale, research expenditures in Category XI have increased greatly. The four distinct subcategories of research into the protection and management of aquatic ecosystems constitute a suite of research activities that were largely absent from the nation's water resources research or were thought of more in the context of water quality research in the 1960s.

It is interesting to note that the trends for the individual subcategories do not always mirror the trend of their major category. For example, although funding for water quality studies (V) declined overall from the mid 1970s to the present, two subcategories within Category V saw modest funding increases—the identification of pollutants (V-A) and understanding the sources and fate of pollutants (V-B).

As far as balance among the 11 major categories, the situation in 2000 has shifted to encompass far more aquatic ecosystems research than in 1973 (see Figure 4-8). Much of this has come at the expense of funding for Categories III (water supply) and V (water quality). In general, however, there is greater parity among the major categories of water resources research in 2000 than there was in 1973.

Federal Agency Budget Breakdown

A complementary way of presenting the data is to consider how individual federal agencies distributed their expenditures among the 11 major modified FCCSET categories. Although many of the individual offices that reported conducting and funding water resources research from 1965 to 1975 have changed (see Table 4-1), most of the cabinet-level federal agencies involved in supporting water resources research in the mid 1960s are still major players in this enterprise today.

Individual agencies have distinct missions and responsibilities, and they differ in their mandated emphasis on fundamental vs. applied research. These

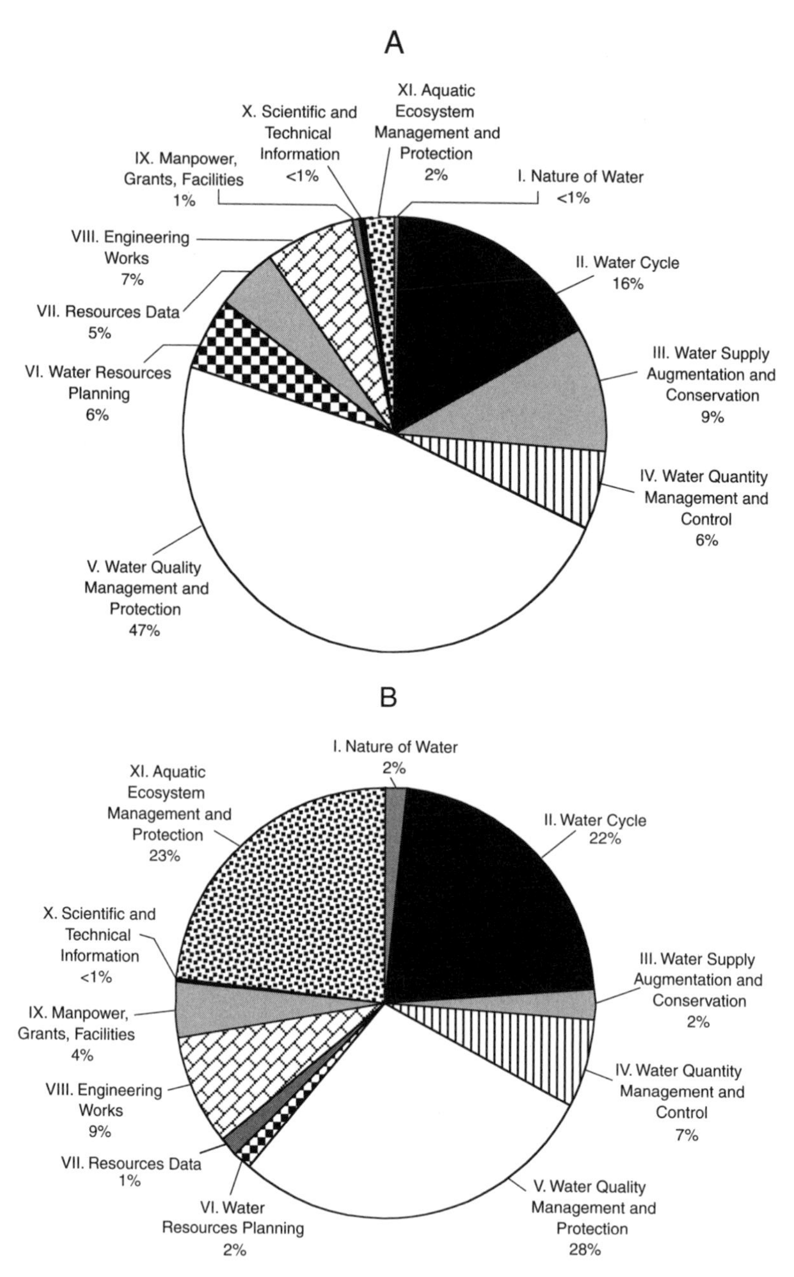

FIGURE 4-8 Percentage of the total federal agency expenditures going to each major FCCSET category for (A) 1973 and (B) 2000. The total funding for water resources research was $705,725,000 in 1973 and $677,971, 000 in 2000. All dollar values are constant FY2000 dollars.

agencies' total budgets and the distribution of their research funding exhibit considerable differences as well. In addition to considering the specific distribution of funding by agency and category, the following section briefly reviews the research focus of each agency. (Although the agencies are listed alphabetically in Tables 4-1, 4-3, and 4-4, the following narratives are ordered by the size of the agencies' funding for water resources research.)

National Science Foundation

Since its inception in 1950, the NSF has awarded grants and contracts that support and strengthen science and engineering research and education, and it plays a significant role in supporting the research infrastructure of the nation's universities, including support of equipment, student fellowships, individual researchers, and multidisciplinary research teams. Although both basic and applied research efforts are included, the NSF has long been considered a primary, if not the primary, agency supporting curiosity-driven research at academic institutions and nonprofit organizations to address fundamental issues. Although much of the benefit to society of this research is expected to accrue over long periods, some benefits occur in a short time frame.

Water-related research is found in nearly all NSF organizational units as well as in crosscutting initiatives such as the Water and Carbon Cycle initiatives, but there is no single program for water resources research. Research supported by the Directorate for Biological Sciences is relevant to the categories of aquatic ecosystems (XI) and the water cycle (II), and it encompasses a network of Long-Term Ecological Research sites (some of which have an aquatic component). The Directorate for Engineering supports research into water and wastewater treatment; the fate, transport, and modeling of contaminants; and sensors and sensor networks for water quality measurement. The Directorate of Geosciences supports traditional hydrologic science, hydrologic–ecological interactions, and meteorological and climate studies related to the water cycle ranging from precipitation processes to long-term trends in water characteristics. The Directorate for Mathematical and Physical Sciences funds water-related research in chemistry and mathematics, including fundamental properties of water and ice, fate and transport of chemicals in water, and mathematical modeling of the water cycle. The Directorate for Social, Behavioral, and Economic Sciences funds a variety of projects that explore fundamental social, economic, cultural, geographic, or decision-related aspects of human activity related to water resources and facilitates international research linkages and experiences for students and investigators. The Office of Polar Programs supports research on aquatic ecosystems in Arctic regions and Antarctica. The Directorate for Education and Human Resources provides funding for water-related education projects across the full range of directorate programs. Support areas include teacher preparation, curriculum development, informal education projects, and digital library resources. NSF also

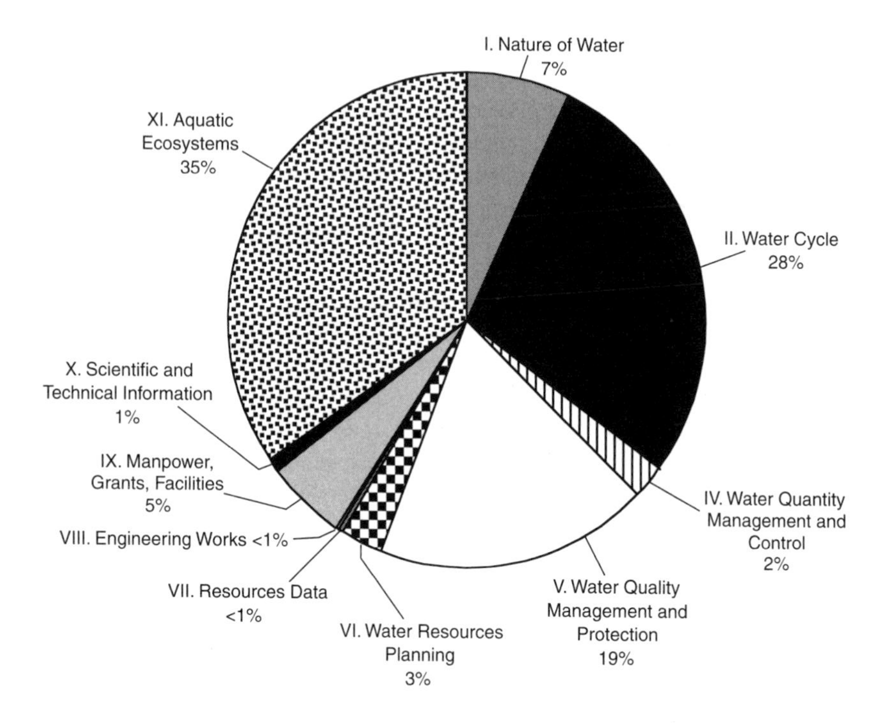

FIGURE 4-9 National Science Foundation FY2000 expenditures by major category ($150,892,000 total). Dollar values reported are constant FY2000 dollars.

supports Science and Technology Centers to address water purification, water resources management in semiarid climates, and river processes.

In FY2000, NSF accounted for 22 percent of total federal water resources research expenditures, four-fifths of which was allocated to categories XI (aquatic ecosystems), II (water cycle), and V (water quality), as shown in Figure 4-9. These results are expected, since these categories deal more with fundamental water processes than with water-related operations, data collection, or administration. In addition, NSF has focused its water resources research on the natural rather than the social sciences, accounting for the low percentage of funds spent in category VI (water resources planning and other institutional issues).

U.S. Geological Survey

Created in 1879, the USGS has evolved from being an agency that performs "surveys" of the nation's land, mineral, and water resources to one that encompasses a broader view of earth science. Its current mission is to provide reliable scientific information to describe and understand the earth; minimize loss of life and property from natural disasters; manage water, biological, energy, and mineral resources; and enhance and protect our quality of life (USGS, 2002). The USGS today is organized into four major disciplines. The Biological Resources Discipline—added to the USGS in 1995—focuses on status and trends of natural systems, basic ecological understanding, analysis of threats, and application of knowledge to management and stewardship. In the past, the Geology Discipline has focused on fundamental geological processes, hazard assessment, and energy and minerals. Its portfolio today includes water-related topics, including the environmental impacts of climate variability, the geological framework for ecosystem structure and function, and geological controls on groundwater resources and hazardous waste isolation. The Geography Discipline is responsible for building, maintaining, and applying *The National Map*. The discipline leads in partnerships with state and local governments and the private sector in producing state-of-the-art geographic tools and products, such as topographic, geological, and hydrographic maps of the entire nation.

The mission of the Water Resources Discipline (WRD) is to "provide reliable, impartial, timely information that is needed to understand the Nation's water resources" (USGS, 2004). As such, USGS scientists conduct research on a wide array of issues central to human and environmental health, including drinking water quantity and quality, impacts of population growth, urbanization and other land-use changes, suitability of aquatic habitat for biota, hydrologic hazards, climate, surface water and groundwater interactions, and hydrologic system management. The WRD maintains a long-term data collection program to collect, manage, and provide scientifically based information that describes the quantity and quality of waters in the nation's streams, lakes, reservoirs, and aquifers. Prime examples include the network of stream gages, the National Stream Quality Accounting Network (NASQAN) program, the Hydrologic Benchmark Network program, and the groundwater-level network. Communication of data and information is a priority of all USGS disciplines. Data collected by the USGS as part of its monitoring activities are widely used by a variety of agencies responsible for water and environmental management and by businesses, citizens, and researchers in government, academia, and the private sector. Trends in funding for data collection per se are discussed in Chapter 5.

With one programmatic exception, all water-related research is conducted in-house by USGS scientists. USGS has several types of programs for conducting research: a centrally coordinated National Research Program in the hydrologic sciences; distributed research investigations, including district offices, the National Water Quality Laboratory Methods Group, portions of the National Water Quality Assessment Program (NAWQA), the Yucca Mountain Project, and the Cascades Volcano Observatory; and programs that are primarily focused on the geology or biology disciplines. About 25 percent of USGS funding for water-related research is managed through the National Research Program. The

USGS manages one external water research program, which is the state Water Resources Research Institutes.

USGS, which contributed 18 percent of the total research budget in FY2000, distributed 80 percent of those funds primarily in Categories XI (aquatic ecosystem management and protection), II (water cycle), and V (water quality management and protection) (see Figure 4-10). Within those three categories, funds were distributed evenly among most subcategories. Aquatic ecosystems research is a growing area of emphasis and is conducted within the Biological Resources and Water Resources disciplines. USGS research supports federal resource management needs in the Everglades, the San Francisco Bay-Delta, the Snake and Columbia river systems, and other systems. Fundamental research on the water cycle and water quality is led by the Water Resources Discipline. The Survey's research on the water cycle has provided a critical foundation for domestic and international hydrologic and energy balance studies related to regional water management and flood control, as well as for understanding the longer-term implications of climate change. Research on all aspects of water

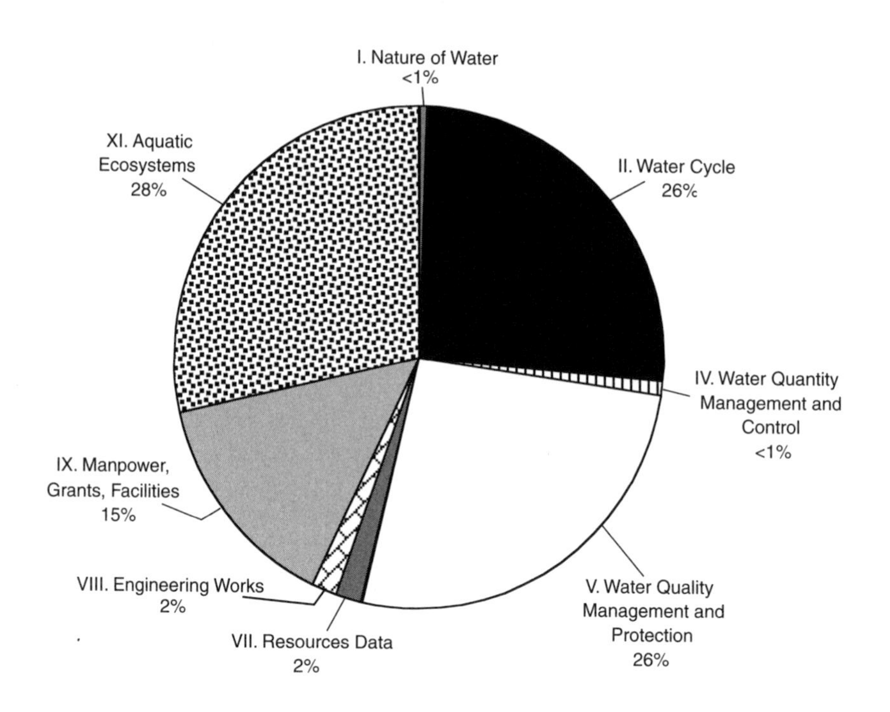

FIGURE 4-10 U.S. Geological Survey FY2000 expenditures by major category ($123,108,000 total). Dollar values reported are constant FY2000 dollars.

quality is driven by priorities and problems identified by state and local partners, the NAWQA program (initiated in 1991), and programs related to toxic chemical and aquatic ecosystem function. Category IX (manpower, grants, facilities) accounted for 15 percent of USGS expenditures, the majority of which supported water research facility infrastructure such as the Hydrologic Instrumentation Facility. Some of these funds, authorized by the Water Resources Research Act of 1965 as amended, provide for research, education, and information exchange with the 54 state Water Resources Research Institutes. In addition, a small part of that program ($1 million) supports a competitive grants program at academic institutions to address high-priority water issues.

U.S. Department of Agriculture

The USDA supports water-related research with four major agencies. The Agricultural Research Service (ARS) is the principal in-house research agency for USDA and contributes around two-thirds of the USDA funding for water resources research. Its research agenda is organized under 22 national programs, one of which is Water Quality and Management, which supports three main programs relevant to the nation's agricultural water resources. The first of these three programs, agricultural watershed management, includes research on water supply and use on irrigated and rain-fed lands to optimize water use and resolve competing demands through establishment of science-based technologies and management. Irrigation and drainage management research, the second program, is intended to support efficiency and sustainability in the face of anticipated declines in water availability. Third, water quality protection and management research emphasizes reducing water contamination from agricultural lands. ARS research is conducted at nine research centers and 83 research locations situated throughout the nation. The missions of several of the research locations including the USDA Water Conservation Lab in Phoenix, Arizona, and the USDA Salinity Lab at Riverside, California, address water and water-related topics. Although most of its research is done by USDA scientists (some of whom hold faculty positions), ARS further involves university faculty in research projects through partnerships and cooperative agreements. Although much of the ARS research is applied (to agricultural problems), there is also a significant component of basic research.

The Cooperative State Research Education and Extension Service (CSREES) provides about 20 percent of the USDA's funding for water resources research. Its primary functions are to identify, develop, and manage programs to support university-based and other institutional research, education, and extension in order to advance knowledge for agriculture, the environment, human health and well-being, and communities. The majority of research functions of CSREES are carried out by faculty at the nation's land grant colleges and universities. In addition to providing core research support for these faculty, CSREES operates a number of annual, extramural research competitions that pertain to the effects of agricultural practices on aquatic ecosystems and watersheds, including nonpoint source nutrients and other contaminants. CSREES activities fall into three program areas: (1) the National Research Initiative, (2) the Hatch Act, and (3) the 406 National Integrated Water Quality Program. Most of the research focuses on water and watershed issues as they relate to agriculture and the conservation of agricultural resources.

The U.S. Forest Service (USFS) has contributed about 18 percent of the USDA funding for water resources research in recent years. It maintains an in-house staff of researchers, including forest hydrologists, fisheries scientists, aquatic ecologists, and climate researchers, to support its management of 155 national forests and 20 national grasslands and to provide outreach to managers of state and private lands. Research units are organized into eight regional research stations, and many of the scientists at the research stations hold academic appointments at nearby universities. USFS scientists cooperate extensively with colleagues from universities and other agencies engaged in forestry and related research and in integrated studies such as the six Long-Term Ecological Research studies that are focused on USFS experimental watersheds. Thus, the research programs of the USFS are closely integrated with and related to the programs of research carried out by faculties at the schools and colleges of forestry. The USFS maintains a broad disciplinary orientation among its research staff, including hydrology, fisheries, aquatic ecology, climatology, engineering, and economics. Research focuses include watershed management, aquatic–terrestrial interactions, fisheries, water quality, and ecology. There are extensive programs of research on watershed management that address issues related to water yields and the maintenance of water quality from forest and range lands. Although much of the USFS research is applied research directed to problems of managing forests and grasslands, the USFS also devotes a substantial effort to investigate basic, long-term questions.

The Economic Research Service (ERS) provides economic analysis relevant to support a competitive agricultural system as well as promote harmony between agriculture and the environment. The general areas of research include the economics and policy dimensions related to agriculture, food, natural resources, and rural development. In the water domain, the agency provides survey-based information on irrigation and water use for agriculture. The ERS economic research program informs USDA policy and program decisions affecting water quality and wetland preservation, and it also derives economic implications of proposed or alternative water quality regulations for the food, agricultural, and rural sectors. The research activities of ERS also address institutional issues such as those related to water transfers and water markets and institutional responses to drought and water scarcity. All of this research is conducted by an internal staff of economists and social scientists, because this agency makes few extramural grants outside of its food, nutrition, and invasive species management program areas. This is the only federal agency having an interest in water research that focuses exclusively on economic and institutional issues, and it constitutes less than 1 percent of the USDA funding total for water resources research.

In FY2000, these agencies within USDA accounted for 17 percent of total federal water resources research expenditures. Research into the water cycle (II), mainly under ARS, accounts for one-third of USDA water-related research, and this research appears to be distributed across multiple aspects of the water cycle in an agricultural context. Water quality management and protection (V), supported by ARS and CSREES, is the next-largest category and includes research aimed at water quality control, pollutant source and fate, and waste disposal. All of the agencies with the exception of ERS support research on water quantity management and control (IV), due to the importance of agriculture in influencing runoff and other watershed processes. Aquatic ecosystem research (XI) at USDA derives almost exclusively from the fisheries research done by the USFS. Water supply augmentation and conservation (III) supported by ARS includes research primarily into water yield improvement and agricultural water use conservation. Together these five categories make up virtually all USDA water-related research expenditures, as shown in Figure 4-11.

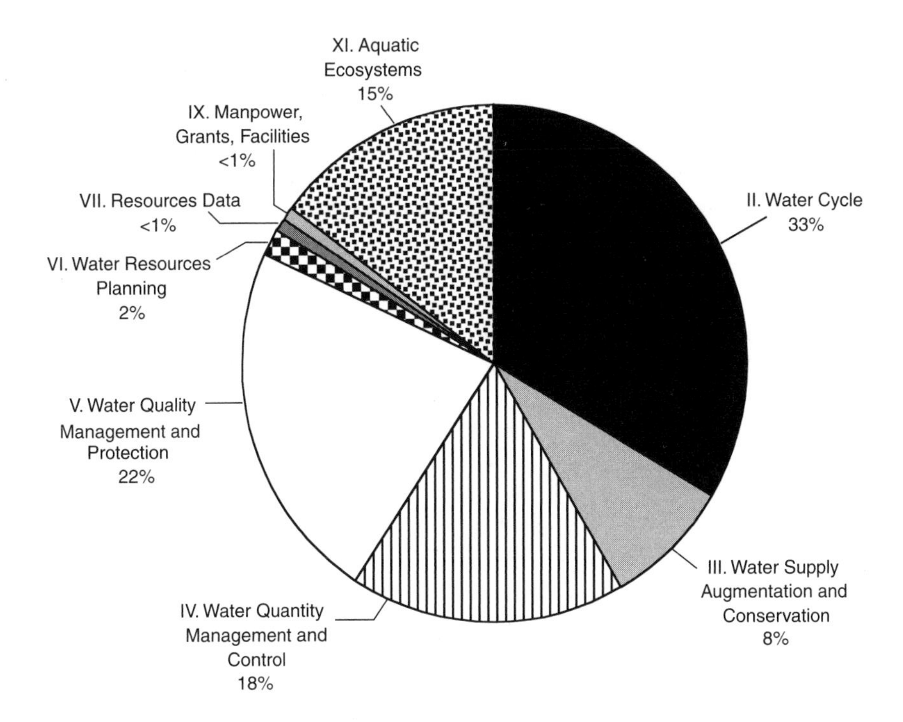

FIGURE 4-11 Department of Agriculture FY2000 expenditures by major category ($116,126,000 total). Dollar values reported are constant FY2000 dollars.

U.S. Department of Defense

About 80 percent of Department of Defense (DoD) funding for water resources research is contributed by the Corps, with lesser amounts coming from the Office of Naval Research (ONR) (about 2 percent) and the Strategic Environmental Research and Development and Environmental Security Technology Certification Programs (SERDP/ESTCP) (about 18 percent). Programmatic areas for the Corps include navigation systems, flood and coastal protection, environmental technologies, infrastructure engineering, geospatial technologies, and integrated technologies for decision making. The mission of the Corps is largely operational in nature, and thus most of the research is highly applied. Its regulatory responsibilities are generally limited to specific provisions of the Clean Water Act, particularly with respect to managing and regulating wetlands. Engineering works, hydraulic modeling, soil mechanics, and contaminated dredge materials are some major research focuses, with the majority of Corps activity being based at seven research laboratories around the nation. In-house research is conducted at seven research laboratories (Coastal and Hydraulics Laboratory, Geotechnical and Structures Laboratory, Information Technology Laboratory, and Environmental Laboratory, all in Vicksburg, MS; Cold Regions Research and Engineering Laboratory, Hanover, NH; Construction Engineering Research Laboratory, Champaign, IL; Topographic Engineering Center, Alexandria, VA) or at the Institute for Water Resources at Fort Belvoir, VA, and its Hydrologic Engineering Center at Davis, CA.

ONR coordinates, executes, and promotes the science and technology programs of the U.S. Navy and Marine Corps through grants to schools, universities, government laboratories, and nonprofit and for-profit organizations. It provides technical advice to the Chief of Naval Operations and the Secretary of the Navy and works with industry to improve technology manufacturing processes. SERDP/ESTCP carries out research to reduce costs and environmental risks by developing cleanup, compliance, conservation, and pollution prevention technologies. In addition, SERDP/ESTCP funds research on cleanup of contaminated defense sites, DoD compliance with environmental laws and regulations, and measures to reduce defense waste streams.

DoD activities account for 15 percent of the overall federal total for water resources research. As shown in Figure 4-12, about half of DoD funds are spent in Category VIII (engineering works), which is anticipated to be necessary to support the Corps' large operational mission, and about a fourth of the funds are spent in Category V (water quality management and protection), which is where almost all SERDP/ESTCP and ONR funds are devoted.

U.S. Environmental Protection Agency

As mentioned in Chapter 2, the EPA was created in 1970 to carry out broad responsibilities for both regulation and research, with an emphasis on protecting

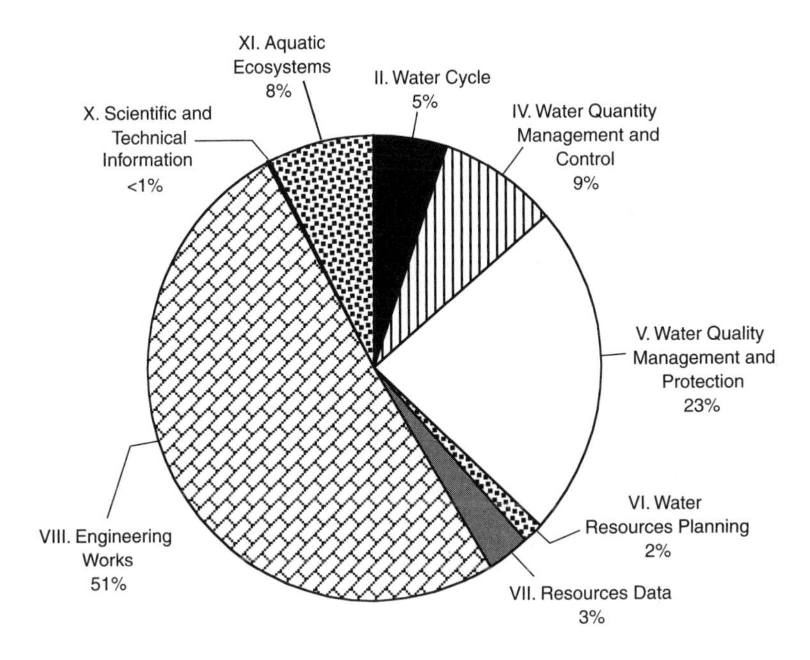

FIGURE 4-12 Department of Defense FY2000 expenditures by major category ($104,668,000 total). Dollar values reported are constant FY2000 dollars.

human and environmental health by protecting our land, air, and water. EPA's Office of Research and Development (ORD) carries out diverse water-related research activities that focus principally on water quality, including microbial pathogens and chemical contaminants and their impact on drinking water and ecosystems; human and ecological health and risk assessment; water quality criteria to support designated uses of fresh waters; tools for assessment, protection, and restoration of impaired aquatic systems; and improved water and wastewater treatment technologies. Water resource-related research is supported primarily through ORD's network of national laboratories and its grant programs.

ORD has five branches that support research centers and laboratories at 13 locations across the country: the National Center for Environmental Assessment (NCEA), the National Center for Environmental Research (NCER), the National Exposure Research Laboratory (NERL), the National Health and Environmental Effects Research Laboratory (NHEERL), and the National Risk Management Research Laboratory (NRMRL). Through NCER, EPA runs competitions for STAR (Science Targeted to Achieve Results) grants and graduate and undergraduate fellowships, provides research contracts under the Small Business Innovative Research Program, and supports other research assistance programs.

NHEERL oversees a network of researchers and facilities, with headquarters in Research Triangle Park, North Carolina; additional laboratories with activities in freshwater resources include the Western Ecology Division (Corvallis, Oregon), the Mid-Continent Ecology Division (Minnesota and Michigan), the Gulf Ecology Division (Gulf Breeze, Florida), and the Atlantic Ecology Division (Narragansett, Rhode Island). Many components of the research conducted by the branches apply to water issues. NRMRL divisions include the Water Supply and Water Resources Division, as well as the Ground Water and Ecosystem Restoration Division. NERL includes the Microbiological and Chemical Exposure Division. EPA primarily emphasizes research relevant to national priorities for safeguarding the environment, although its STAR programs allow investigators to pursue wide-ranging fundamental scientific issues with direct relevance to applications.

EPA's share of total federal water resources research (15 percent in FY2000) is strongly dominated by Categories XI (aquatic ecosystems) and V (water quality), followed by II (water cycle) and IV (water quantity management and control), as shown in Figure 4-13. To a significant extent, research in these categories responds to EPA's regulatory mandates as dictated under the Safe Drinking Water Act, the Clean Water Act, the Comprehensive Environmental

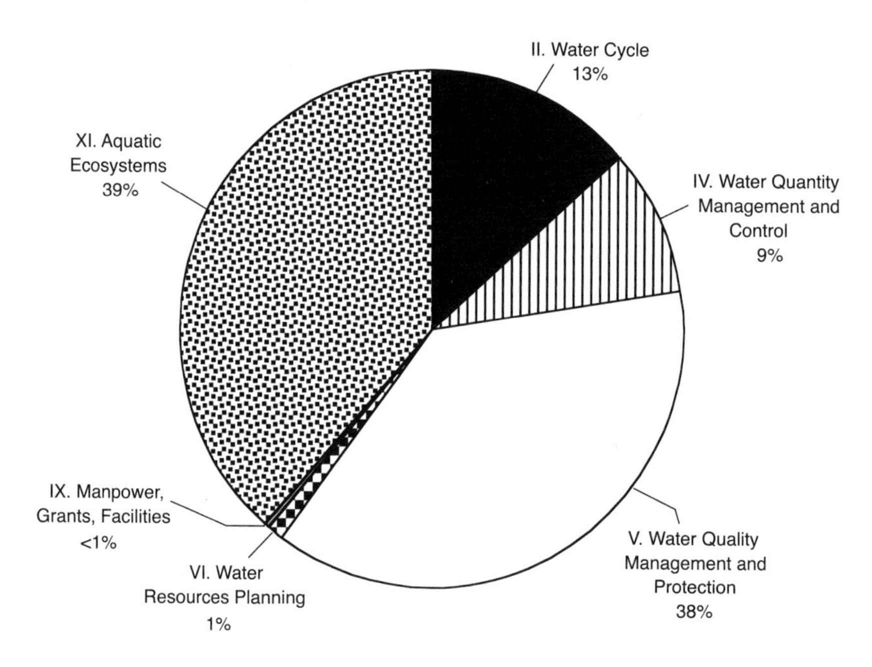

FIGURE 4-13 Environmental Protection Agency FY2000 expenditures by major category ($98,970,000 total). Dollar values reported are constant FY2000 dollars.

Response, Compensation, and Liability Act, and the Resource Conservation and Recovery Act. Funding for water quality management (V), for example, primarily emphasizes water and wastewater treatment and contaminant fate and transport studies.

U.S. Department of Energy

The mission of the DOE is to advance "the national, economic, and energy security of the United States; to promote scientific and technological innovation in support of that mission; and to ensure the environmental cleanup of the national nuclear weapons complex" (DOE, 2003). The nondefense portion of the department is organized into offices specializing in technologies related to specific energy sources such as fossil fuels, nuclear energy, and renewable energy; each office supports applied research in its field. For example, the Office of Fossil Energy has identified three major goals in its water–energy research and development strategy: reduce the use of freshwater resources in fossil energy production and use, improve the quality and reduce the volume of water used in fossil energy production, and reduce water management costs over conventional technology. Its research accounts for around 10 percent of the DOE total for water resources research.

The majority of the DOE funding (about 90 percent) for water resources research comes from the Office of Science, which sponsors studies of the fundamental physical, chemical, and biological processes affecting the fate and transport of contaminants in the subsurface. Much of this research is sponsored by the Environmental Management Science Program, which supports basic research that could enable new, faster, less expensive, and more effective methods for the cleanup of the nuclear weapons complex.

As shown in Figure 4-14, DOE's 3.8 percent share of total federal water resources research expenditures is almost exclusively in Category V (water quality management and protection)—not unexpected given its responsibility for some of the most complex hazardous waste sites in the nation. Most DOE-funded water research relates to pollutants and waste treatment associated with the cleanup of hazardous chemical and radioactive waste at former nuclear weapons production facilities. Other water quality research is directed at the pollutant streams from fossil fuel energy production, affecting both surface water and groundwater.

Apart from the water-related research that was reported to the committee, extensive studies are carried out in other programs, but these studies are not classified as "research" for various statutory reasons and were not included in the survey response. For example, since 1978, DOE's Office of Civilian Radioactive Waste Management has spent "billions of dollars on characterization studies," a significant portion of which was focused on hydrology, climate change, and the fate and transport of radionuclides at the proposed nuclear waste repository site at Yucca Mountain, Nevada (DOE, 2002, p. 33). Similarly, the Office of Environ-

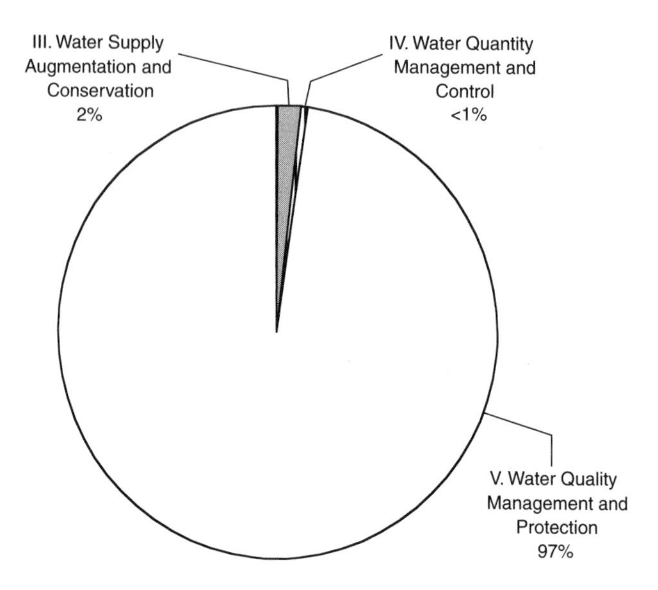

FIGURE 4-14 Department of Energy FY2000 expenditures by major category ($26,053,000 total). Dollar values reported are constant FY2000 dollars.

mental Management has conducted extensive characterization studies of contaminated DOE sites in support of cleanup efforts.

National Oceanic and Atmospheric Administration

Several programs within NOAA (which is within the U.S. Department of Commerce) contribute to water resources research. The National Weather Service (NWS) comprises just less than 10 percent of the reported NOAA total, with the rest split fairly evenly between the National Ocean Service (NOS) and programs within the Office of Oceanic and Atmospheric Research (OOAR). The NOAA water resources research programs have a primarily water cycle and coastal focus, although several other activities concern freshwater resources. (For the purposes of the NOAA survey, "coastal" refers to the land and water area extending from the inland boundary of coastal watersheds to the seaward boundary of the United States Exclusive Economic Zone. In the Great Lakes region, this includes the watersheds of the Great Lakes and St. Lawrence River.) In particular, the NOAA Office of Global Programs (OGP) carries out most of the fundamental research pertaining to the water cycle, to climate predictability and prediction studies and their role in water resources management, and to the social impacts of climate variability and change. The NWS through the Office of Hydrologic Development, Hydrology Laboratory, carries out applied research relating to hydrologic and hydraulic modeling and forecasting at all spatial scales. The Advanced Hydrologic Prediction Service (AHPS) program of the Office of Hydrologic Development provides new hydrologic information and products through the infusion of new science and technology into the operational forecasting process. The inclusion of the Great Lakes within NOAA's Great Lakes Environmental Research Laboratory (GLERL) brings NOAA further into areas of freshwater research. NOAA's coastal research is concerned with agricultural nonpoint source pollution and the resulting hypoxia in coastal waters, as well as other land-based pollutants delivered via runoff. In all the areas, research is conducted both in-house and through grants to external research organizations (universities, nonprofits, and other private sector organizations).

NOAA research accounts for 3.7 percent of the federal total, mainly in Categories II, V, and XI, which focus, respectively, on water cycle processes, water quality management and protection in estuarine and freshwater systems, and aquatic ecosystems (Figure 4-15).

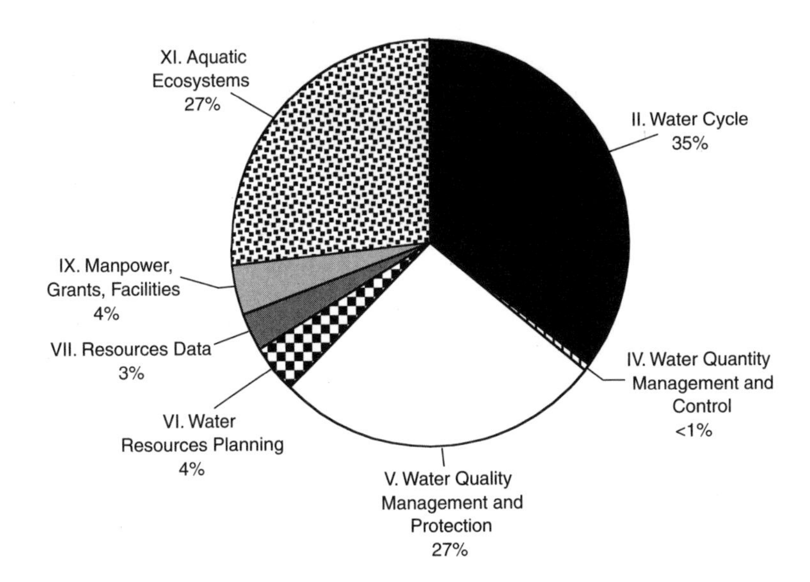

FIGURE 4-15 National Oceanic and Atmospheric Administration FY2000 expenditures by major category ($24,715,000 total). Dollar values reported are constant FY2000 dollars.

U.S. Bureau of Reclamation

The U.S. Bureau of Reclamation (USBR), within the Department of the Interior, is a major supplier of drinking and irrigation water and hydroelectric power. Its mission is confined to the 17 western states. As an operating agency, USBR water resources research is concentrated on applied topics of water development and management. USBR focuses on four main areas: improving water and hydropower infrastructure reliability and efficiency, improving water delivery reliability and efficiency, improving water operations decision support with advanced technologies and models, and enhancing water supply technologies. Examples of research include desalination, river system modeling, fish passage and entrainment, and operational efficiency enhancements.

USBR research accounts for 2 percent of the federal agency total and is distributed across Categories II (water cycle), III (water supply augmentation and conservation), IV (water quantity management and control), V (water quality management and protection), and VIII (engineering works) (see Figure 4-16). Consistent with its mission as a nonregulatory/operating agency, the USBR devotes a large part of its research expenditures to support the operation of water control structures as well as desalination.

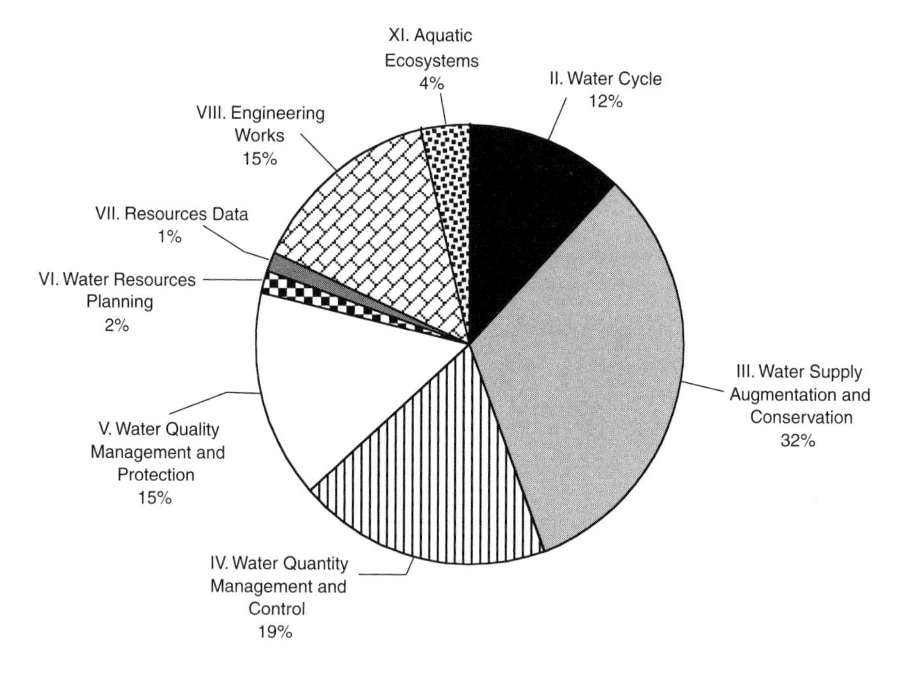

FIGURE 4-16 U.S. Bureau of Reclamation FY2000 expenditures by major category ($14,207,000 total). Dollar values reported are constant FY2000 dollars.

National Aeronautics and Space Administration

Historically, NASA has developed satellite missions in broad support of many different problems related to the earth sciences including environmental change, tectonophysics, oceanography, hydrology, and glaciology. Science teams are formed early in the process of mission design and are involved with converting satellite measurements to useful science products. Once useful information becomes available on a routine basis, NASA provides funding to university and other researchers for innovative application toward enhanced understanding of critical earth science questions. As a result, it is difficult to determine the direct financial influence of NASA programs on hydrologic sciences. Only the more recent satellite missions (e.g., Aqua, Terra, TRMM) have had continental hydrologic sciences as a theme. With most other missions (e.g., TOPEX/Poseidon and Jason-1, Landsat, GRACE), the products support a myriad of other science programs in oceanography, environmental change, agriculture, and forestry. Nevertheless, data from these missions are beneficial for hydrologic sciences, especially for large-scale studies. The NASA estimate of the annual support for water resources research (1.5 percent of the federal total) is primarily directed to a better understanding of fundamental water cycle processes (Category II), as illustrated in Figure 4-17. This includes direct research support to investigators addressing specific hydrologic problems. There has been minor support for studies

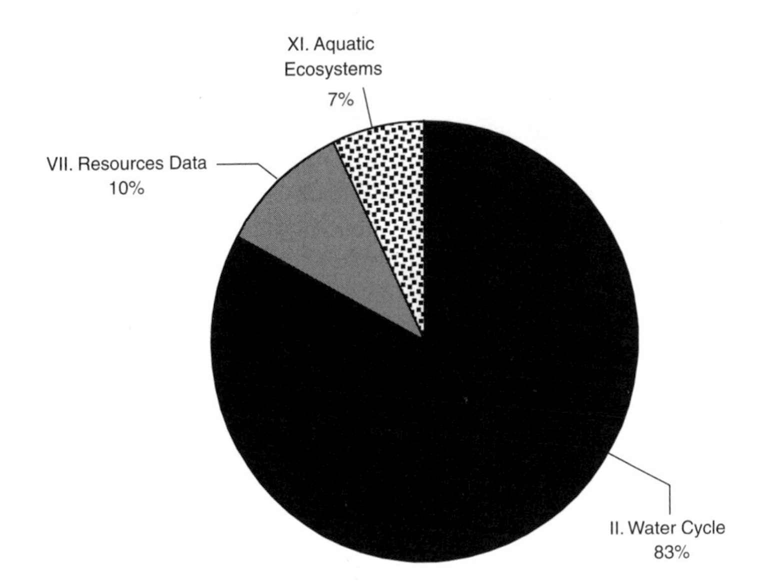

FIGURE 4-17 National Aeronautics and Space Administration FY2000 expenditures by major category ($10,100,000 total). Dollar values reported are constant FY2000 dollars.

involving water quality. Because the total NASA expenditures in satellite development and in operation and funding of science teams for the various missions are of a magnitude that would dwarf all other research expenditures considered in this report, and because the percentage of those activities devoted to water resources research is impossible to determine, they are not included in the NASA survey response.

Department of Health and Human Services

The Department of Health and Human Services (DHHS) has over 300 research and service programs in 11 operating units, eight of which are in the U.S. Public Health Service (PHS). Two major units of the PHS are the National Institutes of Health (NIH) and the Centers for Disease Control and Prevention (CDC). Both NIH and CDC house water-related research activities, but neither has federally mandated regulatory authority for water-related research or management.

NIH's environmentally related research is focused in the National Institutes of Environmental Health Science (NIEHS) and the National Cancer Institute (NCI). The mission of NIEHS centers on reducing morbidity linked to environmental causes. The agency supports research, prevention, intervention, and communication programs. NIEHS includes the National Toxicology Program, which conducts toxicological research; examines reproductive, developmental, cancer, and immunotoxicity outcomes; and develops alternative models. NCI conducts and supports research and its application to prevent, control, detect, diagnose, and treat cancers. The intramural research unit contacted to respond to the survey was the Division of Cancer Epidemiology and Genetics's Occupational and Environmental Epidemiology Branch, which conducts epidemiologic studies to evaluate cancer risks and determines whether they are associated with water contaminants, primarily chemicals. (Thus, the survey does not reflect research conducted or funded by other intramural research programs or extramural grants at the NCI.)

CDC provides health surveillance programs to monitor and prevent disease outbreaks and exposures, conducts research, implements programs and services to prevent disease, and maintains vital statistics and other health databases for the nation. However, CDC neither has funding legislatively directed toward water resources, nor does it have a research program specifically directed toward linking water contaminants and human health outcomes. The agency has recently consolidated with the Agency for Toxic Substances and Disease Registry (ATSDR). ATSDR's mission is focused on preventing hazardous exposures from waste sites and related adverse health outcomes. The agency accomplishes its mission through the conduct of public health assessments, health studies, surveillance activities, health education services, and toxicological profiles of hazardous chemicals. Two programs account for nearly three-quarters of the agency's water-related funding—the Research Program on Exposure-Dose Reconstruction (EDRP) and the Great Lakes Human Health Effects Research Program (GLHHERP). The

EDRP conducts applied research to reconstruct past and model potential levels of contaminants in environmental media and water distribution systems from the source to the receptor populations. The GLHHERP is focused on the 11 critical contaminants identified by the International Joint Commission (United States and Canada). The program characterizes exposures to these contaminants; identifies at-risk populations; investigates the potential for acute and chronic adverse health outcomes; and conducts community-based research, education, and interventions.

The total water resources research funding from DHHS in FY2000 was $9.13 million, all in Category V-H (effects of waterborne pollution on human health), and thus, no figure of the DHHS breakdown is provided. During the three years of the survey period, the NIEHS contribution ranged from 48 percent to 68 percent of the DHHS total, the ATSDR contribution ranged from 18 percent to 51 percent, and the NCI contribution ranged from 1 percent to 14 percent.

Budget Breakdown by Major Modified FCCSET Category

A final level of analysis involves considering which agencies supported the 11 major modified FCCSET categories. Of the 11 major categories, multiple agencies play significant roles in six, while the remaining five are largely within the domain of a single agency. While the data were not reported in terms of the continuum from basic to applied research, nor from the standpoint of external grants vs. in-house agency research, some inferences can be drawn from the liaison reports, the stated agency missions, and the experience of committee members.

As shown in Figure 4-18, research into the nature of water (Category I) is funded primarily by NSF, with a small percentage from the USGS. Most of this funding is expected to support basic research in universities and other research organizations.

Research into the water cycle (II) is the third-largest funding category and is well distributed across the federal agencies (as shown in Figure 4-19). NSF, USDA, and USGS together provide three-quarters of the research funds in this area, but five other agencies, particularly EPA, NOAA, and NASA, also provide research funding. This area likely includes a diversity of subtopics, ranging from fundamental investigations of evapotranspiration and runoff to applied studies in agricultural landscapes. Assuming that the contributions of NSF, as well as some of the funding from USDA and EPA, are through extramural grants, perhaps one-third to one-half of the research occurs at universities and research institutions, and the remainder is conducted by federal agency scientists.

Research into water supply augmentation and conservation (Category III) is largely through the USDA, although USBR contributes about one-third of the total via desalination work (see Figure 4-20).

As shown in Figure 4-21, nearly half of the research into water quantity management and control (IV) is through the USDA, although the EPA and DoD each contribute about 20 percent of the total. This distribution is expected, given

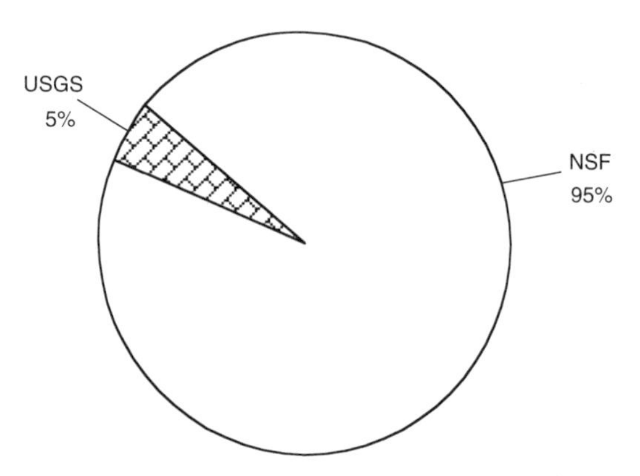

FIGURE 4-18 FY2000 expenditures in Category I (nature of water) by federal agency ($11,153,000 total). Dollar values reported are constant FY2000 dollars.

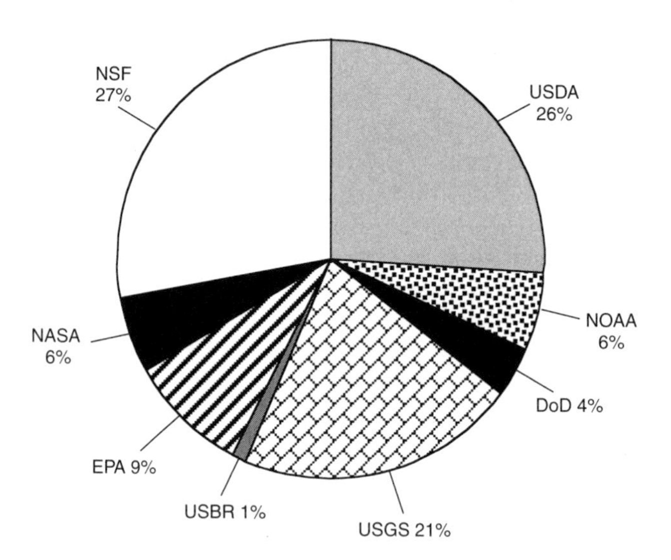

FIGURE 4-19 FY2000 expenditures in Category II (water cycle) by federal agency ($150,835,000 total). Dollar values reported are constant FY2000 dollars.

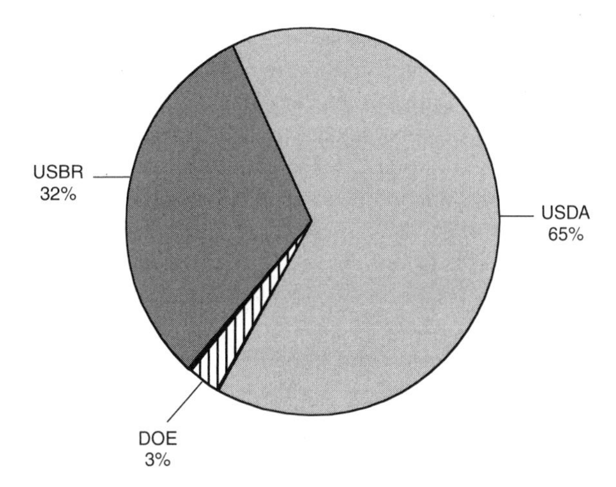

FIGURE 4-20 FY2000 expenditures in Category III (water supply augmentation and conservation) by federal agency ($14,456,000 total). Dollar values reported are constant FY2000 dollars.

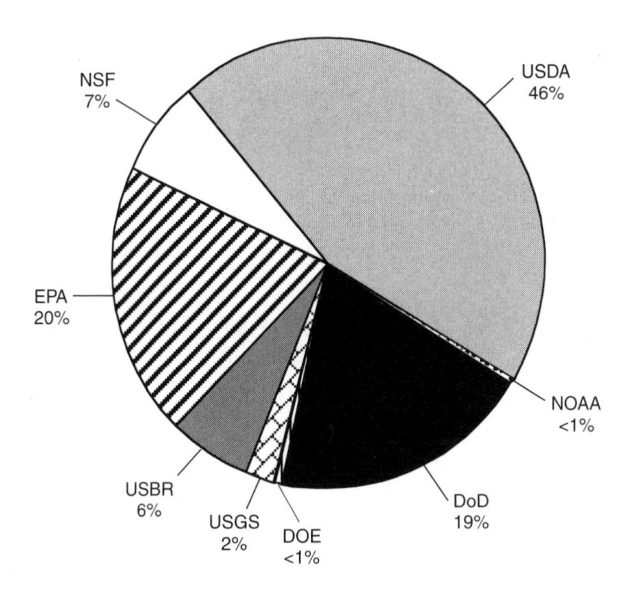

FIGURE 4-21 FY2000 expenditures in Category IV (water quantity management and control) by federal agency ($45,629,000 total). Dollar values reported are constant FY2000 dollars.

the importance of controlling polluted runoff from agricultural lands within USDA and given EPA's interest in understanding the impact of different land uses on surface water and groundwater flow rates. This research activity is presumed to include a mix of agency science, contracts, and external grants.

Research into water quality management and protection (Category V) is the largest single funding category and is arguably the most widely distributed across agencies. Six agencies (EPA, NSF, DoD, USGS, USDA, and DOE) each contribute 13 percent to 19 percent of the total, and three others report 1 percent to 5 percent contributions (see Figure 4-22). All of the water resources research reported by the DHHS falls under the subcategory of understanding the effects of waterborne pollution on human health. However, most of the contributions of the other agencies are widely spread among the eight subcategories. Much of this work likely falls in some intermediate region between basic and applied research, although most of it is likely to be motivated by application. The fraction that takes place in universities and research organizations may exceed one-third, assuming that USDA and EPA each make substantial grants in this area, in addition to NSF.

As shown in Figure 4-23, NSF is the largest single supporter of research into water resource planning (Category VI), followed by USDA and DoD. This cat-

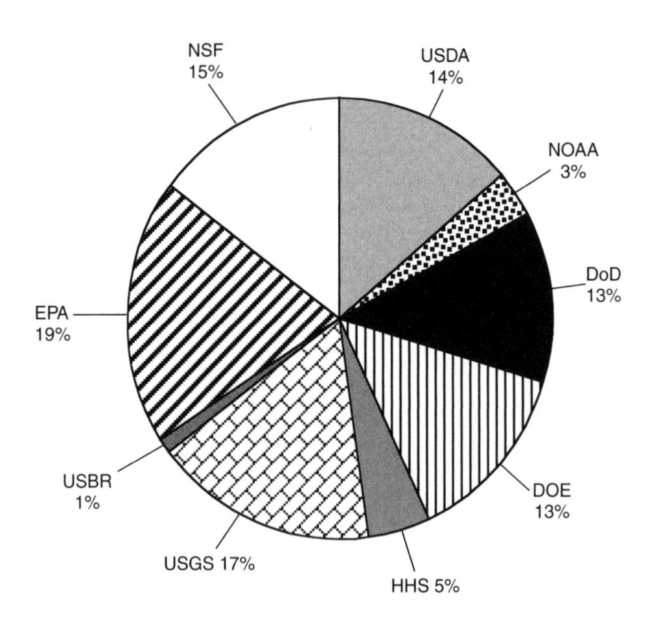

FIGURE 4-22 FY2000 expenditures in Category V (water quality management and protection) by federal agency ($191,669,000 total). Dollar values reported are constant FY2000 dollars.

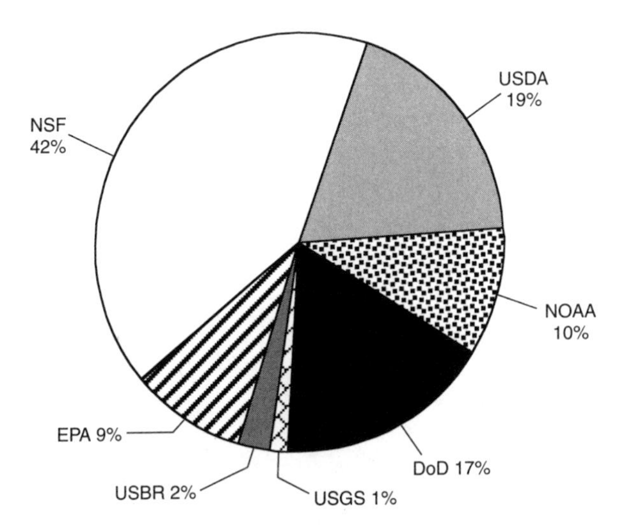

FIGURE 4-23 FY2000 expenditures in Category VI (water resources planning and other institutional issues) by federal agency ($9,834,000 total). Dollar values reported are constant FY2000 dollars.

egory was interpreted to include academic research into decision-related aspects of human activity and other social sciences within NSF, management and planning in an agricultural context within USDA, and integrated technologies for decision making within the Corps. It is likely that somewhat over half of the research occurs in academic or research institutions, and much is intermediate on the continuum of fundamental vs. applied research.

Over half of the research in water resources data (Category VII) is supported by DoD and USGS, although most of the surveyed agencies play a role (Figure 4-24). Research into data acquisition, storage, standards and delivery, modeling, and information technologies are especially prominent topics for the two agencies leading in this area, which serve such diverse needs as integrated assessment modeling, network design, and interagency collaboration. The majority of this research is presumed to be motivated by near-term application needs and to take place within the federal agencies. It is noted again that this category does not encompass expenditures on actual data collection (see Chapter 5 for a more comprehensive discussion of data collection).

Research into water engineering works (Category VIII) is almost exclusively the province of the Corps (and thus DoD) (see Figure 4-25). This is research of a highly applied and practical nature to support the Corps' Civil Works program and presumably is carried out by Corps scientists and engineers and through contracts.

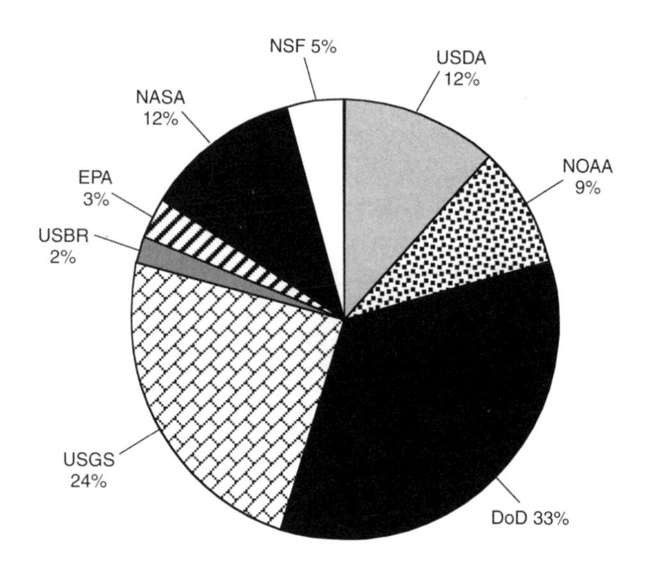

FIGURE 4-24 FY2000 expenditures in Category VII (resources data) by federal agency ($8,679,000 total). Dollar values reported are constant FY2000 dollars.

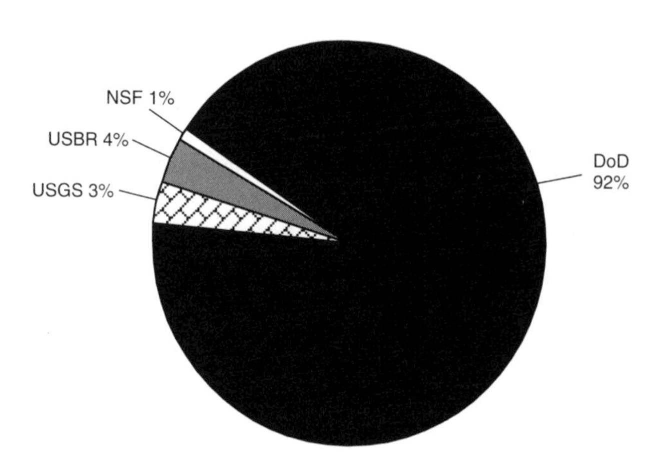

FIGURE 4-25 FY2000 expenditures in Category VIII (engineering works) by federal agency ($58,118,000 total). Dollar values reported are constant FY2000 dollars.

As shown in Figure 4-26, research supporting Category IX (manpower, grants, and facilities) is largely through the USGS, followed by NSF. These expenditures by USGS support facilities. Within NSF, these funds support water-related education projects across the directorates and research facilities.

Support for research into Category X (scientific and technical information) is almost entirely from NSF (Figure 4-27).

Category XI, aquatic ecosystems, is the second-largest individual funding category and includes significant support from NSF, EPA, USGS, and USDA (Figure 4-28). This category was not present in the original FCCSET categories but was developed to reflect emerging research interests since the 1960s in the areas of ecosystem and habitat conservation, ecosystem assessment, climate change, and biogeochemical cycles. NSF supported research in all four subcategories but allocated less to climate change than the other three subcategories, which it supported roughly equally. EPA funding was entirely for aquatic ecosystem assessment. USGS supported primarily the ecosystem conservation and assessment categories. The USDA funding, primarily from USFS, supports watershed and fish habitat research. If funding support by a number of federal agencies is a measure of importance, only Category V enjoys funding as diverse as that of Category XI (compare Figures 4-22 and 4-28). The funding for aquatic ecosystems protection and management appears to support a healthy balance of agency and external researchers and a range of fundamental to applied research.

Nonbudgetary Survey Information

As mentioned earlier, the survey included questions about the federal agencies' missions with respect to water resources (see Box 4-1, question 2), as well as questions about the liaisons' concern irrespective of their agencies' missions (question 5). Although it is not possible to quantify the liaisons' responses to these questions, it is useful to examine the similarities between what the agencies' stated missions are and what the liaisons believe are important emerging water resources issues. The results, presented in Table 4-3, show that few liaisons expressed future issues of concern (column 3) that would not fit into their own agency's research agenda. Furthermore, it is clear that the current agency missions (column 2) do not add up to a national research agenda for water resources.

In terms of emerging issues (column 3 of Table 4-3), there are several commonalities among the agencies' responses. The importance of extreme events and the effects of global climate change; the fate, transport, and effects of pollutants; the nature and control of nonpoint source pollution; and the maintenance and restoration of aquatic ecosystems are issues that emerged from more than one liaison response and are also priorities for this committee, as reflected in Table 3-1. Each of these is a complex, multifaceted problem that reflects many scientific, economic, and societal factors; thus, the overlap is not surprising. However, most agency liaisons phrased their perception of these large-scale issues in terms of

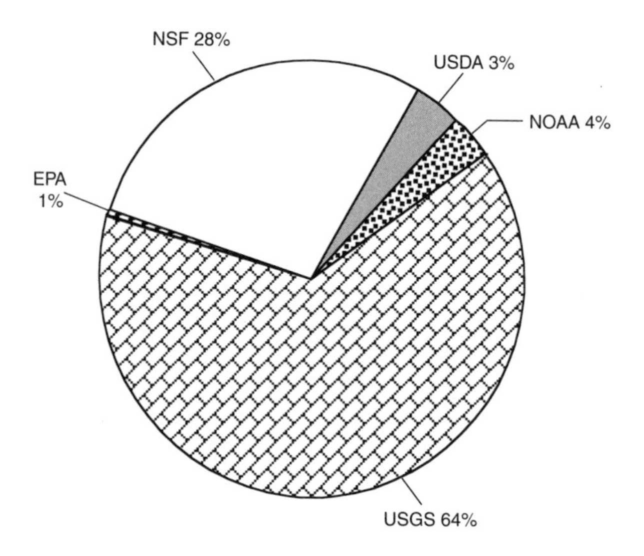

FIGURE 4-26 FY2000 expenditures in Category IX (manpower, grants, and facilities) by federal agency ($27,994,000 total). Dollar values reported are constant FY2000 dollars.

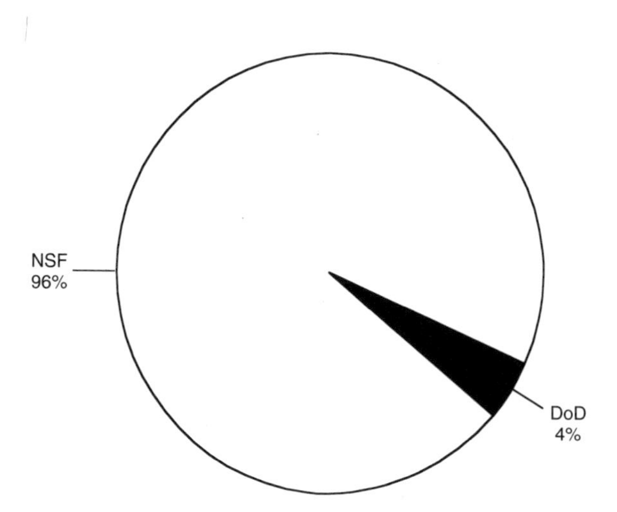

FIGURE 4-27 FY2000 expenditures in Category X (scientific and technical information) by federal agency ($1,168,000 total). Dollar values reported are constant FY2000 dollars.

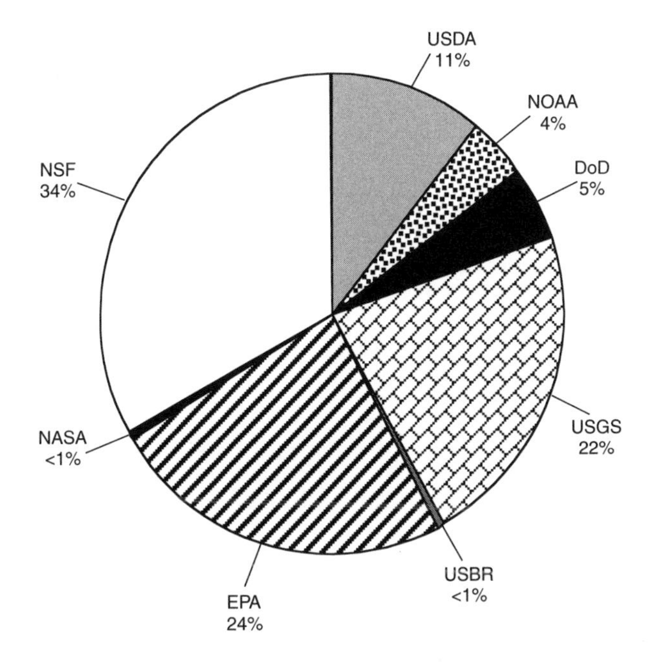

FIGURE 4-28 FY2000 expenditures in Category XI (aquatic ecosystem management and protection) by federal agency ($158,436,000 total). Dollar values reported are constant FY2000 dollars.

their particular missions, which tend to vary widely depending on whether water is the primary focus of the agency mission (e.g., water resources division of USGS) or is a necessary ingredient for some other purpose (e.g., production of food or energy). Thus, it is unlikely that the overlap in the identification of emerging issues is reflected in coordination of research among agencies.

Several additional questions asked of the federal agency and nonfederal organization liaisons during the third committee meeting related to (1) whether the agency's research is conducted extramurally or internally, (2) the time horizon of the research, and (3) whether there are significant place-based studies that are not captured in the submitted budget data. The first question partly reflects whether the research is agency mission-driven or investigator-driven, because research conducted externally is likely to be less constrained by the agency's mission. The time horizon refers to whether the research is expected to have short-term or long-term benefits, which can sometimes be correlated with whether the research is basic or applied (as discussed in Box 3-1). Responses to these questions are summarized in Table 4-4. (It should be noted that not all the agencies responded to these questions, particularly those agencies that submitted their surveys after May 2003.)

TABLE 4-3 Summary of Federal Agency Responses About Their Strategic Plan and Future Water Resources Research Priorities

Agency	Responses to Survey Question 2—Mission
Agriculture	
USDA ARS	Program 201: to develop innovative concepts for determining the movement of water and its associated constituents in agricultural landscapes and watersheds, and to develop new and improved practices, technologies, and strategies to manage the nations' agricultural water resources. Three components: agricultural watershed management, irrigation and drainage, and water quality protection and management.
USDA CSREES	Strategic plan not yet defined, but protection and improvement of water quality and reducing society's reliance on freshwater resources are among the issues being discussed.
USDA ERS	To provide accurate and timely information on (1) the interrelationships between agricultural water availability and agricultural production, (2) the impact of programs and policies that reallocate water resources used by agriculture, (3) the impacts of agricultural production on water quality, and (4) impacts that water quality policies have on the agricultural sector.
USDA USFS	No stated water-specific mission. General USFS mission is to enhance scientific understanding of ecosystems, including human uses, and to support decision making and sustainable management of the nation's forests—through research, inventory, and monitoring.
Commerce	
NOAA NOS	There are 4 facets to the general NOAA mission: (1) protect, restore, and manage coastal and ocean resources through ecosystem-based management, (2) understand climate variability and change, (3) provide weather and water information, and (4) support safe, efficient, and environmentally sound transportation. NOS has responsibilities under each that focus on coastal and estuarine areas.
NOAA NWS	General NOAA mission given above is applicable. More specific to NWS is increased accuracy and lead time of river and stream forecasts. Specific data the NWS collects in support of its forecasting mission include river, reservoir, precipitation and snow measurements, and precipitation data derived from Geospatial Operational Environmental Satellite (GOES) and NEXRAD.

Responses to Survey Question 5—Emerging Issues

- Water scarcity and how that will affect agriculture in the West
- Extreme climate events (floods and droughts)
- Global climate change
- Aging infrastructure
- Stream corridor restoration in agricultural areas
- Improving irrigation and drainage technology
- Antiquated legal systems for water allocation
- Competition to U.S. agriculture in a more global economy
- Fate, transport, and effect of agricultural chemicals in watersheds
- Total maximum daily load development and agricultural best management practices

- Water use efficiency in all sectors of the economy
- Integrating knowledge into water quantity and water quality policy to reduce uncertainty in water resources planning and management

- Control of unregulated sources of pollution (agricultural nonpoint source pollution, for instance)
- Net benefits of alternative reallocations of water across municipal, industrial, agricultural, and in-stream environmental uses
- Transfer of water away from agriculture, while minimizing losses to the sector and rural communities
- Institutions, such as market systems, that could be used to respond to water shortages

- Homeland security and protection of public drinking water sources
- Competition for water between in-stream values (aquatic ecosystems and endangered species) and off-stream users (urban, agriculture, industry)
- Space and time scales of nonpoint source pollution (i.e., the cumulative effects on water quality from multiple activities distributed over a landscape in space and time)

- Climate change and the effects of extremes on water availability and quality
- Ecosystem vulnerability
- Cumulative effects of multiple stressors (both natural and human-induced)
- Prediction of water flows and long-term prediction of water availability
- Developing decision support systems for watershed and river management that can balance competing uses of water

- Improved real-time management of water resources and aquatic ecosystems
- Estimation of streamflow anywhere in a river basin (i.e., ungaged location)
- Increasing density of real-time, automated precipitation networks
- Ensemble data analysis of multisensor precipitation data
- Remote sensing of hydrologic variables: snow cover, soil moisture, river levels, total water storage, air temperature, vegetation, solar insolation, evaporation
- Improved methods for orographic precipitation analysis

continued

TABLE 4-3 Continued

Agency	Responses to Survey Question 2—Mission
NOAA OOAR	Same 4 general NOAA facets, stated above for NOAA NOS. For OOAR in particular, provide research to enable effective management of fisheries, invasive species, and coastal, ocean, and Great Lakes ecosystem health.
Defense	
Corps	Civil Works Program Strategic Plan: (1) provide sustainable development and integrated management of the nation's water resources, (2) repair past environmental degradation and prevent future environmental loss, (3) ensure that operating projects perform in a manner to meet authorized purposes and evolving conditions, (4) reduce vulnerabilities, risks, and losses to the nation and the Army from natural and man-made disasters, including terrorism, and (5) be a world-class public engineering organization. Current emphases of the R&D program include regional sediment management, systems-wide modeling, assessment and restoration technologies, technologies and operational innovations for urban watershed networks, and navigation economic technologies.
SERDP/ ESTCP	SERDP: to resolve environmental concerns in ways that enhance military operations and improve military systems effectiveness, and to support technology and process development in order to reduce operational and life cycle costs associated with environmental cleanup. Examples relate to development of technologies for remediation of groundwater contaminated with heavy metals and development of bioremediation technologies for treatment of nitroaromatic-contaminated soil and groundwater. ESTCP: to demonstrate and validate promising, innovative technologies that target DoD's urgent environmental needs (e.g., characterization and treatment of range contamination and *in situ* remediation of groundwater.
ONR	None with respect to water resources research.

Responses to Survey Question 5—Emerging Issues

- Future fresh water availability and quality, given population growth and redistribution, climate variability, and increasing demands for water
- Better systems and technology for observing precipitation, areal evaporation/ evapotranspiration, soil moisture, groundwater, and snow pack
- Improved precipitation forecasts on daily to seasonal time scales
- Long-term sustainability of water resources
- Understanding the contribution of mountain water reserves
- Development of water use statistics
- Forecasting and managing water under the extreme droughts and floods
- Special problems for water management posed by multijurisdictional watersheds
- Strategies and technologies for securing a water supply in the face of climate variability and change and other water management uncertainties
- Changing poor policies that affect surface waters and groundwater in the United States
- Water and health issues

Unanswered

- Military range sustainability
- Ammonium perchlorate-contaminated groundwater
- Dense nonaqueous phase liquid (DNAPL) characterization and remediation

Unanswered

continued

TABLE 4-3 Continued

Agency	Responses to Survey Question 2—Mission
Energy	
DOE	The three major goals of the Fossil Energy water-energy R&D strategy are: (1) reduce the use of freshwater resources throughout the fossil-energy production and use cycle, (2) improve the quality and reduce the volume of water discharged from fossil energy operations, including coal mining, coal bed methane and oil/natural gas production, and thermoelectric power generation, and (3) reduce water management costs over conventional technology. The Office of Science research program has two primary elements relevant to the types of water resources research: (1) developing new characterization and remediation techniques and (2) developing an improved understanding of contaminant fate and transport.
Health and Human Services	
NIEHS	National Toxicology Program: rodent and other animal-based research and testing to address potential adverse health outcomes in humans from exposure to waterborne chemical and biological contaminants.
NCI	To assess cancer risks from waterborne contaminants.
ATSDR	Research Program on Exposure-Dose Reconstruction goals: to evaluate human health risk from toxic sites and releases and take action in a timely and responsive public health manner. For many site-specific projects: to ascertain the relationship between exposure to toxic substances and disease.
Interior	
USGS	The Water Resources Discipline: provide reliable, impartial, timely information needed to understand the nation's water resources. The National Research Program (NRP): to generate and disseminate knowledge by conducting fundamental and applied research on complex hydrologic problems, to develop techniques and methodology, and to provide scientific leadership in hydrology to the USGS. NRP investigations integrate hydrologic, geological, chemical, climatological, and biological information related to water resources and environmental problems.

Responses to Survey Question 5—Emerging Issues

- Availability of freshwater and competing demands for this limited resource and how future regulations and restrictions will impact the ability of the U.S. to meet increasing demands for low-cost energy and electricity
- Development of new, energy-efficient water purification (treatment and desalination) processes
- Development of better predictive models for the fate and transport of contaminants in groundwater and surface water systems
- Development of a better understanding of the regional effects of global climate change and their impacts on water availability and quality
- Development of a quantitative basis for managing water quality and ecological impacts of storage and conveyance systems

- Consequences of global warming on water supply
- Use of reclaimed water to meet future demand, and associated complications
- Aging of water distribution systems
- Changes in water disinfection methods away from the use of chlorination to chloramination or ozonation

Unanswered

- Issues that affect human exposure to potential and actual contaminants via water resources, and issues that affect fishability, swimmability, and potability of our nation's waters
- Enhance and develop our capacity to assess human health consequences of exposure to hazardous substances both current and historically
- Ensure that appropriate public health intervention strategies are in place especially for at-risk populations
- Ensure that resources are available for the above challenges, while ensuring the nation has the capacity and technology to respond to citizens' requests for assistance with environmental health issues

- Assessing and understanding the processes that control the distribution of contaminants and pathogens at low concentration levels
- Understanding how to manage a river to protect or restore habitat
- Understanding global cycles of water, C, N, P, S, and metals
- Evaluating water resource sustainability for ecological and withdrawal use at regional scales
- To meet the data needs for these research opportunities, we must have long-term data in accessible databases. The data must adhere to well-documented standards to be useful for examining trends and for understanding and differentiating changes due to climate and land-use patterns

continued

TABLE 4-3 Continued

Agency	Responses to Survey Question 2—Mission
USBR	Appropriated funds under the Science and Technology Program are focused on four main areas: (1) improving water and hydropower infrastructure reliability and efficiency, (2) improving water delivery reliability and efficiency, (3) improving water operations decision support with advanced technologies and models, and (4) enhancing water supply technologies.
Other Agencies	
EPA	Conduct leading-edge, sound scientific research to support the protection of human health through the reduction of human exposure to contaminants in drinking water, in fish and shellfish, and in recreational waters and to support the protection of aquatic ecosystems, specifically, the quality of rivers, lakes, and streams and coastal and ocean waters. Furthermore, apply the best available science (i.e., tools, technologies, and information) to support regulations and decision making for current and future environmental and human health hazards related to exposure to contaminants in drinking water, fish, and shellfish, and recreational waters and for the protection of aquatic ecosystems.
NASA	No current strategic plan for water resources research. A water management program plan is under development that will explain how NASA science data products can be used by partnering agencies.
NSF	No formal agency strategic plan for water resources research, although several planning documents address water issues (e.g., Geosciences Beyond 2000, Complex Environmental Systems, Synthesis for Earth, Life and Society in the 21st Century).

Responses to Survey Question 5—Emerging Issues

- Water supply augmentation (e.g., via desalination and water reuse)
- Increase the ability for the coordinated operation of water systems
- Improve water quality, given agricultural return waters and waste disposal
- Methods to better conserve and optimize existing water supplies
- Methods to more economically and reliably modernize outdated water storage and delivery infrastructure
- Aquatic and riparian invasive species
- Water resources data management information system

- Waterborne disease occurrence
- Development of molecular technologies for monitoring and treatment studies
- Research to strengthen CWA/SDWA linkages
- Human population exposure studies to measure the aggregate (total) exposure
- Alternative approaches for screening/prioritizing drinking water chemicals
- Shift focus from point source discharges to nonpoint sources
- More biological indicators for determining aquatic ecosystem condition
- Large-scale processes and activities as determinants of water quality
- Atmospheric deposition and multimedia sources as water quality determinants
- Increase the role of citizen stakeholders in setting watershed management goals and in implementing action programs at the local and watershed levels
- More efficient, more nearly accurate models and methods, and more explicit representation of uncertainties in decision-making processes used by EPA

- Development of remote sensing capabilities of water quality
- Dearth of international observations
- Reinvigorating the groundwater and stream gage monitoring systems

- Better understanding of the physiological effects of waterborne pollutants on aquatic ecosystems via research on the physiology, behavior, genetics, ecology, and evolution of organisms
- Holistic watershed analysis to quantify relationships among catchment characteristics, stream flow, recharge, water quality, and land-use change
- New technology and management methods for water and wastewater treatment, desalination, and detection and removal of low-level health-related contaminants
- Dynamically interactive modeling to enhance understanding of how aquatic systems evolve over time
- Socioeconomic factors that affect individual decisions about water use and the management of water resources by societal institutions
- Linking information about fundamental properties of water systems at the molecular level with properties at watershed and landscape scales
- How to deal with voluminous data, integrate data of differing scales, incorporate real-time data into models, and assess uncertainty in simulations
- Fundamental advances in computational fluid dynamics for use in mathematical modeling of transport in aquatic and other natural systems

TABLE 4-4 Additional Facts about the Federal Agencies Supporting Water Resources Research

Agency	Internal vs. extramural	Significant place-based research	Short-term vs. long-term, basic vs. applied
Agriculture			
USDA ARS	95% internal	None mentioned	50% address long-term problems; 50% address short- and midterm problems
USDA CSREES	100% extramural	None mentioned	25% address long-term basic research issues; 75% address short-term applied research issues
USDA ERS	>99% internal	None mentioned	Research projects are designed to inform policy questions that may arise over a midterm time frame
USDA USFS	>90% internal	None mentioned	40% address long-term problems; 60% address short- and midterm problems
Commerce			
NOAA NOS	70–80% awarded via competitive peer review, of which 70–80% goes to the extramural academic community	None mentioned	About 70% of the proposed payoff is expected in the 3- to 5-year time frame, with the remaining 30% in the 5- to 10-year time frame
NOAA NWS	Approximately 20% of NWS OHD's research funds are for extramural grants (75% of which are extramural research projects via a competitive program)	None mentioned	25% is directed, use-inspired research, and 75% is applied research for improved operations
NOAA OOAR	OGP: 50% spent extramurally; all of the other offices support only internal research	Not answered	OGP has a long-term commitment and annually funds 3-year projects; GLERL has a long-term commitment to research; other lab programs are assembled from 3-year projects

continued

TABLE 4-4 Continued

Agency	Internal vs. extramural	Significant place-based research	Short-term vs. long-term, basic vs. applied
Defense			
Corps	50-60% internal; 40-50% extramural	Five large regional projects occurred during the time frame of the survey, of which the largest two were included in the response	60% is development, 35% is applied research, and 5% is basic research; most projects are composed of multiple short-term focused R&D targets in the 3-year range; the overall program often gets stretched out due to annual funding uncertainties, but the R&D is characteristically short-term in nature
SERDP/ ESTCP	Not answered	Not answered	Not answered
ONR	95% extramural	None	100% basic research with most projects lasting 3–6 years
Energy			
Office of Science	100% extramural	None	100% intermediate- to long-term basic research. Most projects are funded for 3–5 years at a time, and may be renewed several times
Office of Fossil Energy	100% extramural	None	Short- to intermediate-term applied research
Health and Human Services			
NIEHS	Not answered	Not answered	Not answered
NCI	100% internal	None mentioned	The studies range from 3 to 30 years, depending on the need for follow-up; primary role is to conduct scientific investigations that are considered by regulatory agencies when developing policy
ATSDR	25% internal; 75% extramural	None mentioned	ATSDR supports applied research, and most studies are completed in 3–5 years

continued

TABLE 4-4 Continued

Agency	Internal vs. extramural	Significant place-based research	Short-term vs. long-term, basic vs. applied
Interior			
USGS	Almost all is internal except for the Water Resources Research Institutes	None mentioned	Most of the National Research Program is long-term; about 10% of the rest of the water research done in the water discipline would be classified as long-term
USBR	66% internal; 33% extramural	Place-based research may double budget; (two large projects were included in a revised response)	Most of the research has a short-term and applied focus
Other Agencies			
EPA	3 national labs included in survey response: some extramural money, mainly in-house STAR program: all extramural 2 Centers: internal	Region-specific projects are not included in the survey response (e.g., Chesapeake Bay program)	40% of EPA research is problem driven—in particular, that research specifically designed to meet the needs of the water program at EPA
NASA	25-33% internal; 66–75% extramural	None mentioned	According to OMB, 100% of the NASA budget is applied research; there are almost no long-term, in situ projects, although satellites are multiyear projects
NSF	100% extramural	None mentioned	Varies, but much has a long-term focus

From Table 4-4, it is clear that water resources research is conducted both internally and externally across the federal agencies. However, most agencies (or individual offices within the agencies) support only one or the other. For example, within the USGS, research is conducted in-house, and the majority is based at USGS research centers; other activities are located in USGS district offices and cooperative units at public universities. Many of the researchers in the centers and cooperative units also carry academic appointments. Most USBR research is carried out in-house or as part of multiagency site-based projects. The NSF employs only program staff, such that all research activity is conducted by awardees outside the agency. The exceptions are (1) the Corps, where research is about 60 percent in-house and about 40 percent through contractual arrangements with universities, consulting firms, etc.; (2) the NOAA Office of Global Programs; and (3) the DOE, which provides both grants to university-based researchers and direct funding of scientists in national laboratories like Lawrence Berkeley, Lawrence Livermore, and Sandia.

Only the USGS National Research Program and the NSF have a predominantly long-term and basic research focus. Almost without exception, the federal agency liaisons discussed their research in the context of their agency mission, and noted that usable results are expected in a short time frame (within five years), although this was not always achieved.

Finally, the issue of place-based research that would not be captured in the current survey was greatest for the Corps, USBR, and EPA; only EPA was unable to provide budget information for its two largest place-based projects.

Nonfederal Organization Support of Water Resources Research

A variety of nonfederal organizations contribute to water resources research, although only the largest and most obvious organizations were contacted for participation in the committee's survey. WERF provides funding for applied research of importance to its subscribing members, which include utilities and municipalities, environmental engineering and consulting firms, and industrial organizations. Research areas include primarily water quality issues, particularly pollutants, waste treatment, development of the Total Maximum Daily Load process, and various issues related to water quality management and protection. AWWARF coordinates research for a similar set of subscribers and focuses on drinking water quality, infrastructure reliability, efficient customer responsiveness, and environmental leadership. Among environmental nongovernmental organizations, The Nature Conservancy supports water-related research through its Freshwater Initiative, through field office conservation planning, and through fellowships and small grants to researchers at universities and other organizations.

Also included here are summaries of nonfederal research expenditures from four state Water Resources Research Institutes (WRRIs), which together provide a snapshot of expenditures from nonfederal (mainly state) sources. WRRIs are located in each of the 50 states (as well as the District of Columbia, Puerto Rico, the U.S. Virgin Islands, and Guam) generally within the land grant university for that state. In lieu of contacting all of the institutes, the four that have the largest annual funding were asked to show how their nonfederal water resources research dollars are allocated among the modified FCCSET categories. Some of the research supported by states is coordinated with or carried out by the WRRIs, each of which is a federal–state partnership that (1) plans, facilitates, and conducts research to aid the resolution of state and regional water problems, (2) promotes technology transfer and the dissemination and application of research results,

(3) provides for the training of scientists and engineers through their participation in research, and (4) provides for competitive grants to be awarded under the Water Resources Research Act (discussed in Chapter 2). States and their political subdivisions are becoming increasingly involved in supporting research to resolve local/regional issues particularly in the area of aquatic ecosystems restoration and management, with the ongoing Everglades and San Francisco Bay Delta programs being good examples. This research is generally in partnership with one or more federal agencies. Obviously, research funded or conducted by state and local governments but not reported by these four WRRIs (including work done in other states) is not captured in the following analysis. Depending on the state, this research component may be nontrivial, although it is highly unlikely to be of the magnitude of the overall federal investment in water resources research.

Budget data equivalent to that provided by the federal agencies were collected from nonfederal organizations that support water resources research. However, because the responding organizations were not queried during the earlier FCCSET exercise, no trends analysis can be done showing how this contribution has evolved over the last 30 years. As shown in Figure 4-29, the level of funding for water resources research from nonfederal organizations is very small in comparison to that of the federal agencies—about 5.5 percent of the combined federal and nonfederal total in FY2000. This fraction, and the nonfederal spending for individual major categories, was quite stable over the three years covered by the survey.

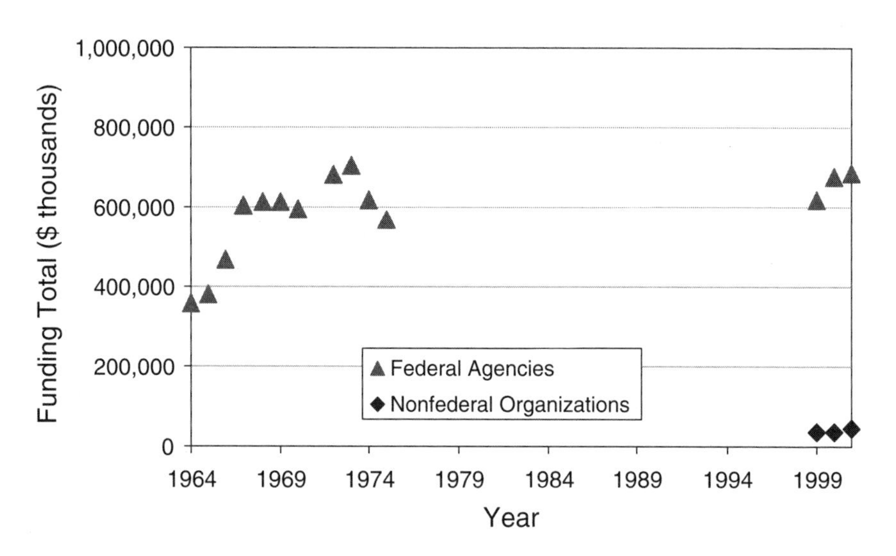

FIGURE 4-29 Expenditures on water resources research, 1964–2001, by federal and nonfederal organizations. Dollar values reported are constant FY2000 dollars.

In terms of which major categories receive the most funds, the nonfederal organizations that were queried mirror the federal agencies in supporting primarily Categories V (water quality) and XI (aquatic ecosystems) (see Figure 4-30), but to an even greater extent. Categories V and XI consume 37 percent and 29 percent of the nonfederal expenditures, respectively, but only 28 percent and 23 percent of the federal expenditures. There are some other interesting differences. The nonfederal organizations provide support for some categories that are clearly lower funding priorities for the federal agencies, notably Categories III (water supply augmentation) and VI (water resources planning and other institutional issues). Alternatively, nonfederal funding for Categories II (water cycle) and IV (water quantity management) is much less, as a percentage of total funding, than the corresponding federal contribution.

The expenditures from WERF, AWWARF, and the four largest WRRIs were substantially different, in terms of the categories of supported research. WERF and AWWARF funds are dominated by water quality (Category V) research (82 percent and 73 percent, respectively), which is not surprising given the stated missions of the organizations (which focus on wastewater and drinking water, respectively) and their constituencies (primarily wastewater and drinking water

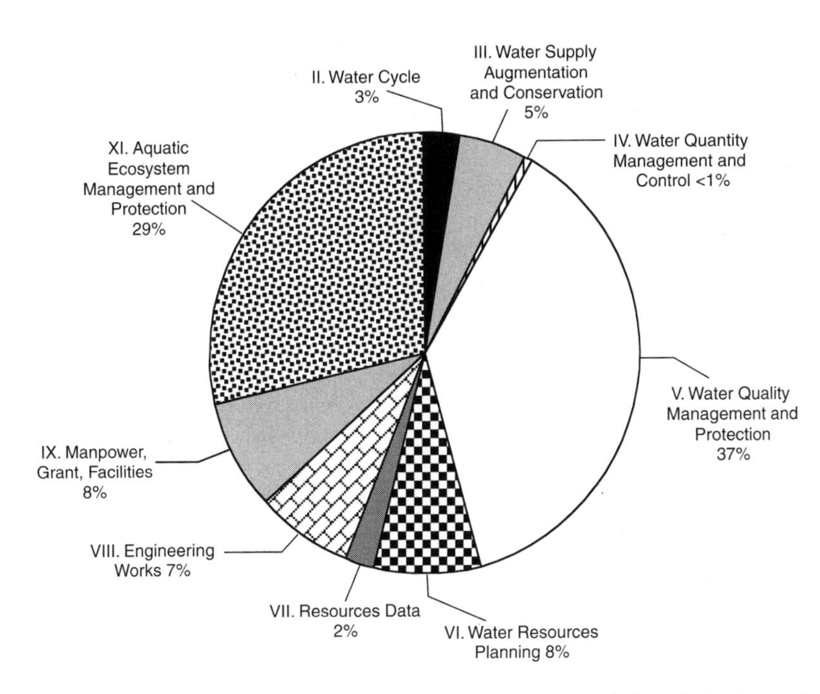

FIGURE 4-30 Percentage of the total expenditures from the queried nonfederal organizations going to each major category for FY2000.

treatment plant operators). All of the research supported by these two organizations is conducted externally by universities, consultants, and other contractors.

The WRRIs, on the other hand, cover the major categories much more uniformly. The Division of Hydrologic Sciences of the Desert Research Institute, University of Nevada, has received around $250,000–400,000 annually for a wide range of projects covering seven of the 11 major categories, with water quality control (Subcategory V-G) being the most funded subcategory. The Pennsylvania Water Resources Research Center and Institutes of Environment reported $5 million to $8 million annually in nonfederal funding for water resources research, although the committee had difficulty distinguishing research expenditures from other expenditures, including staff and infrastructure support. The Texas Water Resources Institute reported nearly $5 million in nonfederal expenditures in 2001, their most complete year of reporting. Water supply augmentation (Category III) and manpower (Category IX) were the dominant categories, and liaison notes indicate these were federal earmarks (although they are unlikely to have been reported by any of the federal agencies). The Utah Water Resources Institute reported annual expenditures of about $5 million, primarily in Categories II, V, VI, VII, and XI.

It is important to note the limitations of the budget data received from the significant nonfederal organizations. As suggested above, the values may have a significant degree of error, if activities other than research were reported or if federal funds were reported. For example, WRRI responsibilities extend to teaching (fellowships) and outreach, not all of which may have been teased out of the reported numbers. Furthermore, as stated earlier it is likely that the total amount of funded water resources research from nonfederal entities is an underestimate, given the small number of participants in the survey (although the survey was sent to all those organizations known to focus on water and to conduct at least $3 million of research annually). In addition, the survey does not reflect very recent trends in topical funding at the nonfederal organizations. For example, since September 11, 2001, there has been a substantial shift in emphasis at AWWARF to fund research related to water security issues. This work was mandated in the 2002 Bioterrorism Act and is being conducted in partnership with EPA. Over $2.35 million is slated for about a dozen water security research projects, each in the $150,000 range, to begin in the 2002–2004 time frame. It is important to keep these limitations in mind when considering both the scope and the magnitude of the nonfederal investment in water resources research.

EVALUATION OF THE CURRENT INVESTMENT IN WATER RESOURCES RESEARCH

Chapter 3 discussed several criteria for evaluating individual research areas for inclusion in the national water resources research agenda, including the national significance of the research, how well research in certain areas has progressed, and the need to have a balanced portfolio of research topics. In the spirit of those criteria, the following evaluation of the current investment in water resources research considers three primary issues. First, the total funding for water resources research and the trends in that funding over the last 30 years are considered, with comparison to other research areas such as national defense, health, and transportation. Second, the specific topical areas that are currently funded are compared to the research areas discussed in Chapter 3 (Table 3-1) as the highest priorities for the next 10–15 years. Finally, the balance of the current research enterprise is evaluated with respect to several factors mentioned in Chapter 3, including whether the research is short-term or long-term, fundamental or applied, investigator- or mission-driven, and internally or externally conducted.

Overall Funding

The level of federal investment in water resources research has not grown from the early 1970s. Statistically stated, there is low to no likelihood that the 1999–2001 total budget values are higher than the 1973–1975 values. Indeed, when Category XI (aquatic ecosystems) is removed from the survey results, one finds that the federal investment has declined, in constant FY2000 dollars, from a high of $691 million in 1973 to $526 million in 2001—a conclusion supported by the uncertainty analysis presented in Appendix C.

One way to evaluate the current federal investment relative to future needs is to compare the growth rate of water resources research funding to economic and demographic parameters such as population growth, the annual gross domestic product (GDP), the annual federal budget outlay, and federal expenditures on water and wastewater infrastructure. While water resources research funding has remained stagnant in real terms over the last 30 years, population has grown from 212 million in 1973 to 285 million in 2001, a 26 percent increase. Thus, the per capita spending on water resources research has fallen from $3.33 in 1973 to $2.40 in 2001. Similar trends are observed when comparing water resources research funding to GDP and the federal budget. GDP has grown steadily for the last 30 years, more than doubling between 1973 and 2001. During that time, water resources research funding has decreased by over half from 0.0156 percent of the GDP to 0.0068 percent. Outlays from the federal budget have increased from $877.2 billion in 1973 (in FY2000 dollars) to $1.857 trillion in 2001, with the portion of the budget devoted to water resources research shrinking by more than half from 0.08 percent to 0.037 percent. All of these trends are shown in Figure 4-31, which plots the 1973–1975 average data vs. 1999–2001 average data for water resources research per capita and for water resources research as a percentage of GDP and of the budget outlay.

The frequency of conflicts surrounding water resources has increased with population growth, most notably in areas where water demands press hard on available supplies (see the five bolded questions throughout Chapter 1). If one

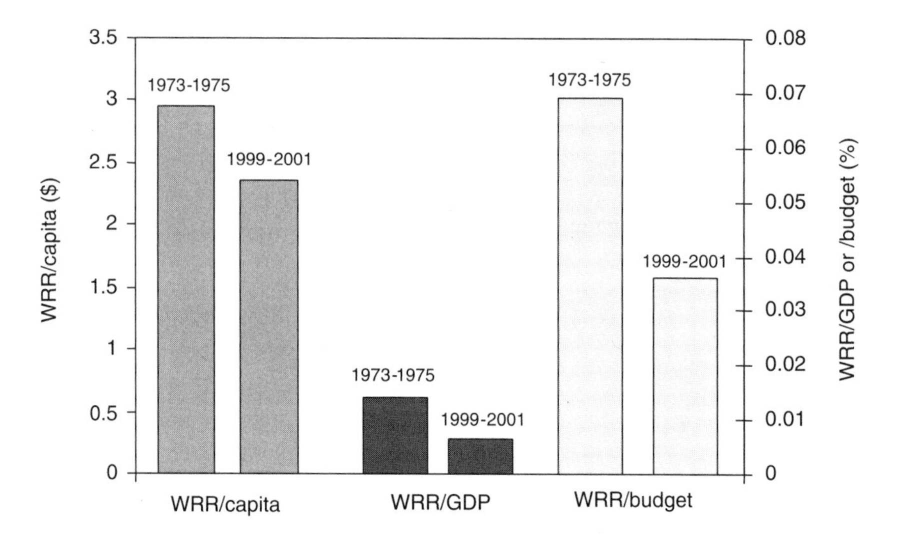

FIGURE 4-31 Bar graph showing how funding for water resources research (WRR) has decreased between 1973–1975 and 1999–2001. Three parameters are shown: water resources research funding per capita, water resources research as a percentage of GDP, and water resources research as a percentage of the total budget outlay. All dollar values in constant FY2000 dollars. Note the different y-axes. SOURCE: Population data from the U.S. Census, GDP data from the Bureau of Economic Analysis, federal budget outlays from the U.S. Government Printing Office (2003).

assumes, as the evidence in Chapter 1 suggests, that the need for water resources research should roughly parallel population and economic growth and the associated increase in conflicts, then the three trends above suggest that current levels of investment in water resources research are insufficient to address future problems. Research is not all that is falling behind on a per capita and per GDP basis. Federal expenditures on water infrastructure projects (drinking water and wastewater only) over the last decade have been stagnant as well (GAO, 2001), despite increasing calls for repair and replacement of aging systems (e.g., GAO, 2002).

Spending on water resources research can also be compared to spending on other lines of research of national importance. Annual spending for research and development across the federal enterprise is compiled by the American Association for the Advancement of Science (AAAS) for 11 functional categories, as listed in Table 4-5. Because of the broad nature of these categories, this information cannot be used to determine research spending in water disciplines, although water issues can be envisioned to fall under several of these categories, particularly environment, agriculture, and general science. Table 4-5 suggests that the current annual expenditure of about $700 million for water resources research pales in comparison to the annual federal support for defense and health, and it

TABLE 4-5 Major Functional Categories of Federal Government Research and Development (values reported in millions of dollars, adjusted to FY2000 dollars)

Category	FY1999	FY2000	FY2001
Defense	43,499	43,160	44,925
Health	16,983	18,758	21,134
Space	8,819	8,437	8,704
General science	5,521	5,593	6,177
Environment	2,090	2,082	2,079
Transportation	1,852	1,664	1,608
Agriculture	1,483	1,561	1,776
Energy	1,217	1,146	1,284
Commerce	506	530	455
International	196	200	245
All other	698	637	616
Total R&D	*82,864*	*83,769*	*89,003*

SOURCES: AAAS (2000, 2001, 2002).

also lags behind expenditures for transportation, agriculture, and energy research—areas that many might agree are comparable in importance to the provision of clean water. Indeed, the difference in the amount of federal support received by the health and water resources fields in FY2000 suggests that health research is 28 times more important than research on water and wastewater services (including all the services provided by aquatic ecosystems). Furthermore, there is evidence that spending in some these other fields has paralleled population and economic growth, unlike water resources research. Figure 4-32 shows data from AAAS on annual federal expenditures for (A) nondefense research and development and (B) combined defense and nondefense research and development. Data on expenditures for health research show remarkable increases, particularly in the last 10 years. This growth in funding for health research has exceeded population growth, such that per capita spending on health increased from $28 in 1973 to $67 in 2001 (both in FY2003 dollars).

Topical Areas

It is also important to analyze the modified FCCSET categories in which the current water resources research funds are being invested, and to compare the results to the areas of research felt to be important over the next 10–15 years. The NRC (2001) report outlined 43 areas of water resources research of paramount importance (see Chapter 3 and Table 3-1). Table 4-6 lists these research areas, and it denotes the modified FCCSET subcategory in which such research would logically fall.

A

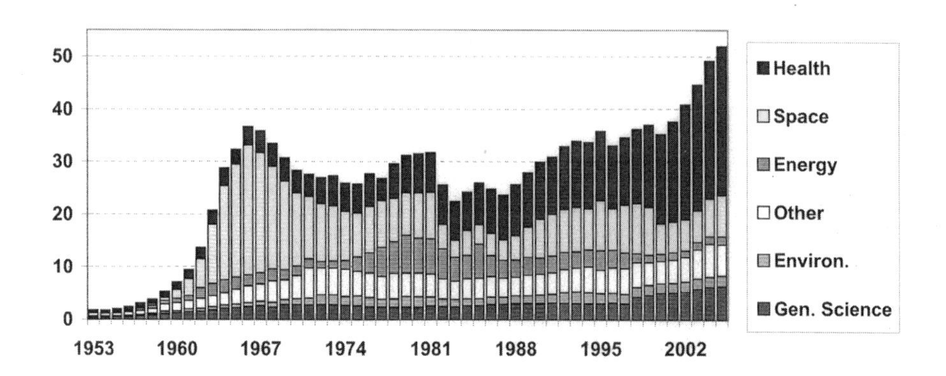

B

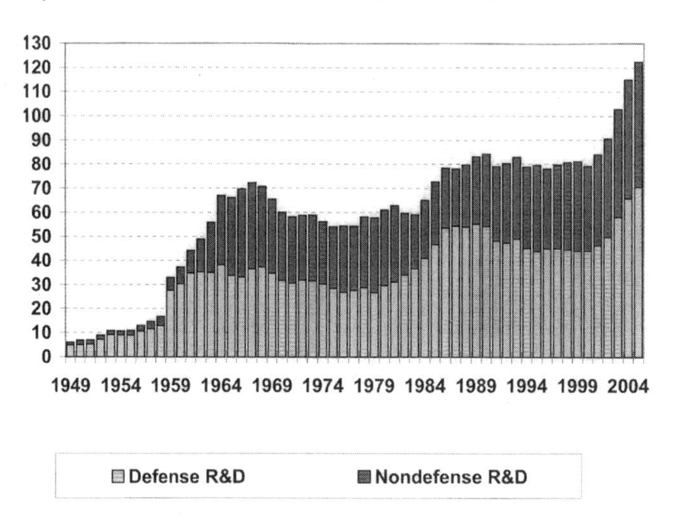

FIGURE 4-32 Trends in (A) nondefense federal research and development and (B) defense plus nondefense federal research and development over the last 50 years. SOURCE: http://www.aaas.org/spp/rd/guihist.htm. Reprinted, with permission, from AAAS (2004). © 2004 by American Association for the Advancement of Science.

TABLE 4-6 Overlap Between the 43 Research Areas in NRC (2001) and the FCCSET Subcategories

Research Area from *Envisioning the Agenda for Water Resources Research in the Twenty-first Century*	FCCSET Category
Water Availability	
1. Develop new and innovative supply enhancing technologies	III-A,B,C; IV-B; less III-D,E,F
2. Improve existing supply enhancing technologies such as wastewater treatment, desalting, and groundwater banking	III-A,B,C; IV-B; V-D,F
3. Increase safety of wastewater treated for reuse as drinking water	V-D
4. Develop innovative techniques for preventing pollution	Small part of V-G
5. Understand physical, chemical, and microbial contaminant fate and transport	V-B
6. Control nonpoint source pollutants	IV-A,D; V-G; VI-F
7. Understand impact of land use changes and best management practices on pollutant loading to waters	II-J; IV-C; V-C
8. Understand impact of contaminants on ecosystem services, biotic indices, and higher organisms	V-C
9. Understand assimilation capacity of the environment and time course of recovery following contamination	XI-A; less V-C
10. Improve integrity of drinking water distribution systems	VIII-J; less V-F
11. Improve scientific bases for risk assessment and risk management with regard to water quality	V-H for humans; V-C for ecorisk
12. Understand national hydrologic measurement needs and develop a program that will provide these measurements	VII-B
13. Develop new techniques for measuring water flows and water quality, including remote sensing and *in situ* techniques.	VII-B; less II-A; V-A
14. Develop data collection and distribution in near real time for improved forecasting and water resources operations	VII-B; less VII-A
15. Improve forecasting the hydrological water cycle over a range of time scales and on a regional basis	II-A
16. Understand and predict the frequency and cause of severe weather (floods and droughts)	II-B,E
17. Understand recent increases in damages from floods and droughts	II-B
18. Understand global change and its hydrologic impacts	II-M
Water Use	
19. Understand determinants of water use in the agricultural, domestic, commercial, public, and industrial sectors	VI-D
20. Understand relationships between agricultural water use and climate, crop type, and water application rates	III-F
21. In all sectors, develop more efficient water use and optimize the economic return for the water used.	III-D,E,F

continued

TABLE 4-6 Continued

Research Area from *Envisioning the Agenda for Water Resources Research in the Twenty-first Century*	FCCSET Category
22. Develop improved crop varieties for use in dryland agriculture	III-F
23. Understand water-related aspects of the sustainability of irrigated agriculture	III-F
24. Understand behavior of aquatic ecosystems in a broad, systematic context, including their water requirements	XI-A
25. Enhance and restore species diversity in aquatic ecosystems	XI-A,B
26. Improve manipulation of water quality and quantity parameters to maintain and enhance aquatic habitats	XI-A
27. Understand interrelationship between aquatic and terrestrial ecosystems to support watershed management	XI-A; less IV-A
Water Institutions	
28. Develop legal regimes that promote groundwater management and conjunctive use of surface water and groundwater	VI-E
29. Understand issues related to the governance of water where it has common pool and public good attributes	VI-E
30. Understand uncertainties attending to Native American water rights and other federal reserved rights	VI-E
31. Improve equity in existing water management laws	VI-E
32. Conduct comparative studies of water laws and institutions	VI-E
33. Develop adaptive management	VI-A; less VI-I
34. Develop new methods for estimating the value of nonmarketed attributes of water resources	VI-C
35. Explore use of economic institutions to protect common pool and pure public good values related to water resources	VI-C
36. Develop efficient markets and market-like arrangements for water	VI-C
37. Understand role of prices, pricing structures, and the price elasticity of water demand	VI-C
38. Understand role of the private sector in achieving efficient provision of water and wastewater services	VI-C
39. Understand key factors that affect water-related risk communication and decision processes	VI-G
40. Understand user-organized institutions for water distribution, such as cooperatives, special districts, mutual companies	VI-E
41. Develop different processes for obtaining stakeholder input in making of water policies and plans	VI-H
42. Understand cultural and ethical factors associated with water use	VI-H
43. Conduct *ex post* research to evaluate the strengths and weaknesses of past water policies and projects	VI-I; less VI-B

Severely Underfunded Research Areas

Several things are immediately obvious from Table 4-6. First, there are many research areas (#28–43) that fall primarily under the heading "water institutions" and are part of Category VI—which, as shown in Figures 4-6 and 4-7, receives a small and declining proportion of the water resources research budget. That is, 100 percent of the "water institution" topics are underfunded. The NRC (2001) report argues vigorously for the need to support research in legal, economic, social, and other less technical areas given the complexity of water resources problems likely to characterize the beginning of the new century. The lack of investment in institutional research is readily explainable. Critical reviews of institutional arrangements and the development of innovative institutions do not fit squarely into the missions of any of the existing agencies that conduct and sponsor water resources research. Inasmuch as institutional research is clearly not a priority with these agencies, they hire very few personnel with institutional expertise and therefore have little capacity to accomplish institutional research. The USGS, for example, has only one economist. Moreover, the agencies involved in water resources research are themselves part of the institutional landscape, and as such they are reluctant to engage in research that might threaten their own standing, mission, or mode of organization. The history of federal agencies is not rife with instances of critical self-examination, and change has come slowly and with difficulty.

There is also an unfortunate tendency among policy makers to believe that institutional research infringes on their policy-making authorities, with the consequence that institutional research tends not to be supported. This is unfortunate because it arises from a general failure to understand that institutional research does not a priori entail the making of policy. Rather, like other scientific research, one of its purposes is to inform the making of policy. As a result, it is rare that any federal agency sees any advantage or incentive to support research on institutions, resulting in very minimal investment and a corresponding lack of innovation and change in the institutional arena. Indeed, the country continues to struggle with water laws and institutions devised to address the problems of the 18th and 19th centuries. It will become increasingly difficult to manage water scarcity and deteriorating water quality (see Chapter 1) with institutions that were designed for other purposes.

Similarly, better understanding of water use in various sectors of the economy (#19) is also felt to be of critical importance to solving future water conflicts. Unfortunately, research on water demand has been ignored, largely because no federal agency has a history of doing demand management. Rather, the focus at the federal agencies has been on water supply. Furthermore, there is little incentive for water suppliers to reduce demand, especially if this cuts into revenues needed to cover costs.

All these research priorities (#19, #28–43), while generally supported in the 1960s during a period of overall growth in funding for water issues, receive only 1.5 percent of the current water budget (see Category VI in Figure 4-8B). Furthermore, this is likely an overestimate of the funding in these particular research areas, since the modified FCCSET categories are significantly broader than the 43 research areas.

To a lesser extent, the same can be said of research topics #20–23 in Table 4-6, which deal with making all sectors, but particularly agriculture, more water efficient. Research in Category III has declined since its height in the late 1960s, when desalination research made up the bulk of the investment in water supply augmentation and conservation issues. Similarly, every subcategory of III, including conservation in domestic and municipal, industrial, and agricultural water use, has declined in real terms since the 1960s. This fact also suggests highly inadequate funding for research areas #1 and 2 in Table 4-6, which concern the creation of water supply-enhancing technologies. These trends are unlikely to change in the near future. Agricultural water users are unlikely to support research that could lead to reductions in the allocations of water that they receive. Despite the national trend away from supply augmentation as a means of resolving water scarcity, agricultural users are for the most part wedded to the notion of protecting existing supplies and augmenting supply as a general strategy for managing scarcity. The federal government has not invested extensively in research on supply-enhancing technology, preferring to leave such investment to the private sector where returns can be fully captured. Thus, for example, the vast majority of investment in desalting and water treatment and purification technologies is in the private sector.

Two other research priorities from Table 4-6 are clearly in need of greater support if future water supply problems are to be averted. First, topic #10 suggests that many of our current drinking water systems are nearing the end of their usable lifetimes, requiring research into their rehabilitation and replacement. As shown in Figure 4-7H, around $3 million annually was devoted to this research topic (subcategory VIII-J) in FY1999–2001, which ranges from only 6 percent to 9 percent of the total budget of Category VIII. (This subcategory was not included in the earlier FCCSET survey, preventing a trends analysis). Research areas #12–14, which deal with hydrologic measurement needs, are encompassed by subcategories VII-A and VII-B, and are also underfunded relative to their importance in solving future water resources problems. As shown in Figure 4-7G, funding for research on new methods for hydrologic data acquisition has fallen by 75 percent since the mid 1970s, while funds for network design have been cut by two-thirds in the same time period.

On the basis of these analyses, it is concluded that more than half of the 43 water resources research priorities noted in Chapter 3 are currently grossly underfunded. A continuation of past funding trends will result in substantial underinvestment in a majority of the areas that have been identified as high-priority areas for the future. Clearly, the patterns of future investment in water resources research will have to be rebalanced if future priorities are to be addressed adequately.

Better Funded Research Areas

Research priorities #3–8, #11, and #15–18 in Table 4-6 fall into FCCSET Categories II (water cycle) and V (water quality) and would appear to be receiving more appropriate absolute levels of funding relative to their importance. Indeed, Categories II and V received a combined 50 percent of all water resources research funding in FY2000, and almost all of the subcategories that overlap with these research priorities are among the best funded in their major category. It should be noted, however, that whether the specific projects noted in Table 4-6 (such as #15—forecasting the hydrologic water cycle over a range of time scales and on a regional basis) are being funded is unclear given the broad nature of some of the subcategories. The need to rebalance the pattern of investment mentioned in the section above suggests that total water resources research funding levels should increase if the absolute level of funding for Categories II and V is to be maintained.

Funding for Category XI, the protection and management of aquatic ecosystems, makes up nearly one-fourth of total water-related FY2000 research expenditures, providing confidence that research areas #9 and #24–27 are receiving support. In general, one way to evaluate the funding for Category XI is to consider the societal risks and potential costs associated with the problems researched. Conflicts between human and environmental uses of water are increasingly costly—the estimated price tags for restoration of the Everglades and the San Francisco Bay Delta (in the $10 billion range over perhaps 20 years) easily top the list, but less costly conflicts appear virtually everywhere and not just in the arid and semiarid West. Future climate change and altered biogeochemical cycles further threaten the health of aquatic ecosystems, and as long as society values healthy, functioning ecosystems and the goods and services they provide, the costs of managing and restoring aquatic ecosystems will be heavy. In this light the current research expenditures appear appropriate. Moreover, the estimated value of healthy ecosystems in providing clean water is rarely precisely known, but in the well-publicized case of New York City's water supply, $300 million invested in ecosystem protection via riparian land acquisition and other activities may save several billion dollars that otherwise would be needed for water purification infrastructure.

A second line of support for continued funding of Category XI comes from the agency liaisons, who were asked to identify the major water issues confronting the nation in the next five to ten years, irrespective of their agency's mission. The perspectives of the USGS, NSF, and EPA are particularly germane, as these agencies provide 80 percent of the funding for Category XI and over half of total water-related research spending. Three of the four topics identified by USGS would fall under Category XI (how to manage a river to restore or protect habitat, understanding global cycles, and evaluating water resource sustainability for ecological and withdrawal use). NSF identified a wide range of fundamental issues, including holistic watershed analysis and enhanced understanding of aquatic ecosystems. EPA's strategic plan for water resources research includes a portfolio of research focused on human and environmental health.

Budget expenditures within the four XI subcategories indicate distinct agency emphasis on and a preponderance of funding for ecosystem fundamentals and assessment. Support of research into biogeochemical cycles should perhaps be considered for augmentation, and research into the ecological consequences of climate change for freshwater systems currently receives less funding than do other subcategories of XI. The observed dramatic increase in funding for aquatic ecosystem protection and management appears to reflect a societal desire to maintain healthy aquatic ecosystems. This of course in no way negates the importance of safe drinking water or adequate supplies for agriculture, but it does indicate a recognition within funding agencies that society is willing to support the high costs of current efforts to minimize harm to aquatic ecosystems and provide for their repair.

Lower-Priority Research Areas

Finally, there are obviously many subcategories of water resources research that are funded by the federal agencies but are not mentioned in Table 4-6 as being priorities. These include almost all of the subcategories in Categories I (nature of water), VIII (engineering works), IX (manpower, grants, and facilities), and X (scientific and technical information). These make up only 14.5 percent of the total water resources research budget. Categories VIII, IX, and X in particular support continued operations of water resources infrastructure, education, and information dissemination—activities that would not have been noted in NRC (2001) as research priorities given the report's topical focus. These subcategories' absence in Table 4-6 should not be interpreted as a suggestion for further reducing their funding.

Appropriate Mix

Assessment of the nation's portfolio in water-related research requires examining existing funding according to multiple and complex criteria. First, as discussed in Chapter 3, this should be partly based on the balance between research that is long-term vs. short-term, fundamental vs. applied, investigator-driven vs. mission-driven, and internal (agency scientists) vs. external (universities, contractors). Furthermore, NRC (2001) identified 43 research priorities within three categories—water availability, water use, and water institutions—and stressed the importance of addressing these three broad areas. Lastly, research activities that incorporate one or more of the four themes presented in Chapter 3 (interdisciplinarity, broad systems context, uncertainty, and the importance of human and ecosystem adaptation) are likely to be most effective in providing solutions to society's most critical water-related problems (see Chapter 1).

Short-term vs. Long-term Research

The terms "long-term vs. short-term," "fundamental vs. applied," "investigator-driven vs. mission-driven," and "internal vs. external" are defined in Box 3-1, which suggests that there are often positive correlations between research that is longer-term, more fundamental, investigator-driven, and externally conducted. Thus, the following analysis does not separately address each of these characteristics, but rather tries to make generalizations based on information from the federal agency liaisons about whether their research programs are short- or long-term and about how much research is conducted internally vs. externally (see Table 4-4).

It is tempting to try to determine the percentage of the total water resources research budget that goes toward long-term vs. short-term research by considering the nature of the individual modified FCCSET categories. Unfortunately, several of the largest categories include research that falls along the entire spectrum from short- to long-term and that addresses poorly understood phenomena or processes that are relevant to applied water issues but require new knowledge. Category II (water cycle) exemplifies research where a basic understanding of processes such as evapotranspiration and runoff is critical to management of a water supply, and where changing land use and climate add new complexities. Category V (water quality management and protection) likewise involves new understanding of contaminant transport, fate, and effect, and draws significant support from NSF and USGS as well as EPA. Categories VI (planning) and VII (resources data), both small slices of the overall funding, are supported by a diversity of agencies. Category XI includes research that is relatively long-term and applied, such as studies of aquatic habitat and the development of assessment methods; it also includes knowledge-generating research motivated by the need for aquatic ecosystem protection and management.

Several other categories clearly are primarily applied and require answers in the shortest practical time frame. Category III (water supply augmentation and conservation), funded primarily by USDA and USBR, is concerned with water supplies for agriculture, urban consumption, and power generation. Category IV (water quantity management and control), dominated by USDA followed by EPA and DoD, emphasizes practical applications to, for example, agricultural watershed management and control of polluted runoff. Category VIII (engineering works) is funded almost entirely by the Corps. Research in these three categories is usually carried out internally or through highly targeted contracts and grants.

Together these categories make up 17 percent of total water resources research expenditures.

Long-term, basic research predominates in few categories. Research into Category I (nature of water) is very fundamental, receives minimal funding (almost exclusively from NSF), and arguably is not an urgent priority. Some elements of Category XI, particularly NSF and USGS support of climate and biogeochemical research, should be considered fundamental research. In these categories, research can be characterized as primarily long-term and basic, and it appears likely that half or more occurs at universities and other research institutions. Despite all these inferences, it is difficult to conclude that a certain percentage of water resources research is short-term vs. long-term based on the modified FCCSET categories alone.

A different approach relies on information provided by the five largest contributing federal agencies regarding the percentage of their funding that is long-term and basic vs. short-term and applied. NSF, USGS, EPA, DoD, and USDA contributed 88 percent of the federal water resources research budget in 2000. The information they provided about the percentage of their research that is short-term vs. long-term and internal vs. extramural (Table 4-4) suggests that at least one-fourth but less than one-half of water resources research reported by the federal agencies can be classified as research of a more basic and long-term nature, likely taking place at universities and nonfederal research institutions. In the view of the committee, between one-third and one-half of the total water resources research portfolio should be allocated to longer-term, more fundamental, investigator-driven research to ensure that critical knowledge will be available on which to base water resources management in the next 10–15 years. Given the current balance, this seems highly achievable with a relatively minor change in emphasis. Considering the emerging problems of contaminants and pathogens that occur at low concentrations, the challenge of reconciling the water demands of humans and ecosystems, the uncertainties associated with climate change and human alteration of biogeochemical cycles, and the pending exhaustion of surface water supplies, there is reason for concern about whether the existing portfolio can provide the needed critical knowledge.

There are several reasons why not enough of the current research portfolio is focused on long-term and fundamental research. OMB examiners explained to committee members that they are often not inclined to support long-term and fundamental research because it does not have immediately usable or useful results and it is hard to judge its effectiveness. Furthermore, the structure of incentives to the agencies tends to be linked to the time scales of elected officials, with the result that there is little emphasis on long-term research. OMB typically supports this outcome by requiring that agencies stick to their stated missions (which may or may not include long-term goals). It seems apparent then that the structure of incentives at the federal level contains some bias against longer-term research on topics in water resources, which likely permeates the development of longer-term research agendas within the federal agencies (although there are some exceptions).

While it is true that the desire for institutional commitment, continuity in personnel, and effectiveness of control favors internally conducted research over externally conducted research, the disadvantages of temporal bias and reduced flexibility should be considered. Research needed to adapt to new conditions might be more difficult for an internally focused agency because resources are tied up in maintaining existing research staff, who may not have the necessary expertise in new and innovative fields. The increased involvement of academic scientists through peer-reviewed grants would strengthen the overall water resources research capacity by expanding the pool of researchers, by providing more flexible research capacity, by placing greater emphasis on competitive evaluation of projects in terms of national priorities and quality of research, and by providing avenues through which the bias against longer-term research can be counteracted. (It should be noted that some federal agencies conduct peer review of internally conducted research projects.)

Water Availability, Water Use, and Water Institutions

In terms of whether the research portfolio adequately addresses the three broad areas of water availability, water use, and water institutions outlined in NRC (2001), both water use and water institutions are currently underfunded. For reasons discussed above, almost every research priority listed under water institutions belongs to a modified FCCSET category that has seen declining proportions of the federal water resources research budget since the 1960s. Those research priorities falling under water use have been similarly neglected with the exception of those within Category XI. While it is certainly not appropriate to suggest that each of the three broad areas of water availability, water use, and water institutions should receive one-third of the annual budget, given that the 43 research areas are of varying breadth and complexity, it is clear that the current distribution is out of balance. The committee does not believe it to be unreasonable that 10 percent of the total water resources research budget be allocated to combined water use and institutional topics. Currently, and as discussed in detail in NRC (2001), almost nothing in known about the determinants and extent of public water uses, and very little is known about water use in other sectors. As discussed previously, institutional topics are similarly under-studied.

Interdisciplinarity, Broad Systems Context, Uncertainty, and Adaptation

One of the categories experiencing a severe reduction in research funding from the 1973–1975 levels is Category VI—water resources planning and other institutional issues (see Table 4-2). Ironically, this is the synthesis category that would support research questions with a strong *interdisciplinary* nature and with a *broad systems context*—key themes for the national water resources research agenda proposed in Chapter 3. This is also the category that would best encompass the significant present-day issues of human and ecological *adaptation* to changing water resources conditions. As argued in Chapter 3, for multiple research agencies to be able to tackle complex emerging water problems in a way that will enhance our understanding beyond an incremental level, the themes of interdisciplinarity, broad systems context, uncertainty, and adaptation must permeate future research in water resources. Although there are a few initiatives that address interdisciplinarity (see Box 4-2 for an example), an increase in research funding for Category VI with specific emphasis on the aforementioned four themes is certainly warranted.

BOX 4-2
Interdisciplinary Initiatives Among the Federal Agencies

Federal agencies have markedly increased their emphasis on interdisciplinary research through new program initiatives, including single-agency calls for proposals and multiagency partnerships. Several NSF initiatives provide instructive examples of progress toward integration across disciplines as well as an increased emphasis on synthesis science and complex systems.

NSF's Long-Term Ecological Research (LTER) network is perhaps the longest-running experiment in interdisciplinary ecological research. Now 20 years old and consisting of a network of 24 sites across diverse ecosystems (including two urban sites), the LTER program has brought together ecologists, ecosystem scientists, hydrologists, geochemists, and other specialists and has led to extensive multidisciplinary collaboration directed at understanding ecosystems and their response to human activities. In FY2002 the LTER program had a funding level of $17.8 million and it supported 1,100 scientists and students (NSF, 2002); the next ten years are expected to emphasize "synthesis science" and incorporate social science more than has previously been the case. Most LTER-funded research has addressed scientific questions best described as fundamental and long-term.

The Water and Watersheds program, beginning in 1995, funded interdisciplinary projects that synthesized physicochemical, biological, and social science expertise in addressing water and watershed issues. NSF,

CONCLUSIONS AND RECOMMENDATIONS

Analysis of the survey budget data and narratives shows that the independent research efforts conducted by various federal agencies to respond to funding mandates of the past only partially recognize the emerging water problems of the future. One clear deficiency is the low and declining funding of research related to water institutions and planning, despite the important role of the social sciences in providing knowledge to help meet future demands for water for human and environmental uses. The current national investment in water resources research needs adjustments in its magnitude and mix to meet the challenges that lie ahead.

A quantitative analysis showed that real levels of total spending for water resources research have remained relatively constant (around $700 million

EPA, and USDA jointly provided funding over Water and Watersheds' six years of existence. A number of the funded studies utilized hydrologic models, GIS land-use analysis, and field studies of physical, chemical, and biological response variables to address topics such as nutrient runoff and ecological status and trends. Those familiar with this program generally saw it as successful, although perhaps not in its involvement of social scientists. Furthermore, the program lacked sufficient support within the agencies for its continuation. Most research activities were motivated by applied issues, but required new knowledge and approaches.

NSF's Biocomplexity Initiative, begun in 1999, emphasizes an interdisciplinary, complex systems approach to environmental research in several areas, of which the programs in coupled human and natural systems and coupled biogeochemical cycles are most relevant to water resources research. A recently established NSF Advisory Committee for Environmental Research and Education indicates that support for this initiative may extend for another decade or more, and it has expressed the need for long-term, well-defined programmatic initiatives in order to incorporate interdisciplinary research and address complex environmental questions and problems (NSF, 2002). Biocomplexity in the environment now encompasses a broad spectrum of NSF-funded research and in 2003 received about $30 million in total research awards. Most of the research can be characterized as fundamental in nature, but with well-articulated relevance to environmental concerns.

in 2000 dollars) since the mid 1970s. When Category XI (aquatic ecosystems) is subtracted from the total funding, there is very high likelihood that the funding level has actually declined over the last 30 years, even under assumptions of significant data uncertainty. In particular, it is almost certain that funds in Categories III (water supply augmentation and conservation), V (water quality management and protection), VI (water resources planning and institutional issues), and VII (resources data) have declined severely since the mid 1970s. Of particular note is the severe reduction in funding for Category VI (water resources planning and institutional issues). Although the historical data and those collected as part of the present survey contain significant uncertainty, this was accounted for using a likelihood framework as described in Appendix C.

Water resources research funding has not paralleled growth in demographic and economic parameters such as population, GDP, or budget outlays (unlike research in other fields such as health). Since 1973, the population of the United States has increased by 26 percent, the GDP and federal budget outlays have more than doubled, and federal funding for all research and development has almost doubled, while funding for water resources research has remained stagnant. This suggests that water resources research has not been accorded priority over the past 30 years. Given that the pressure on water resources varies more or less directly with population and economic growth, and given sharp and intensifying increases in conflicts over water, a new and expanded commitment will have to be made to water resources research if the nation is to be successful in addressing its water and water management problems over the next 10–15 years.

The topical balance of the federal water resources research portfolio has changed since the 1965–1975 period, such that the present balance appears to be inconsistent with current priorities (as outlined in Chapter 3). Research on water demand, water law, and other institutional topics as well as on water supply augmentation and conservation now garners a significantly smaller proportion of the total water research funding than it did 30 years ago. In an absolute sense these categories appear to be significantly underfunded. When the current water resources research enterprise is compared with the list of research priorities for the future, it becomes clear that significant new investment must be made in these categories of research if the national water agenda is to be addressed adequately.

Additional funds should be invested in high-priority topical areas that are currently neglected, including water supply augmentation and conservation, monitoring, and several institutional topics. If enhanced funding to support research in these categories is not diverted from other categories (which may also have priority), the total water resources research budget will have to be enhanced.

The current water resources research portfolio appears heavily weighted in favor of short-term research. This is not surprising in view of the de-emphasis of long-term research in the portfolios of federal agencies. It is important to emphasize that long-term research forms the foundation for short-term research in the future. A mechanism should be developed to ensure that long-term research accounts for one-third to one-half of the portfolio.

OMB should develop guidance to agencies on reporting water resources research by topical categories. Understanding the full and multiple dimensions of the federal investment in water resources research is critical to making judgments about adequacy. In spite of clearly stated OMB definitions of research, agencies report research activity unevenly and inconsistently. In its discussions with federal agency representatives, the committee learned that agencies fund research through multiple budget accounts. Only projects that are specifically funded through research accounts are counted as such and are reported to OMB as research activities. Research funded through operational or "place-based" projects such as the Everglades or the San Francisco Bay Delta is not reported to OMB as research. Failure to fully account for all research activity undermines efforts by the administration and Congress to understand the level and distribution of water resources research. This problem could be remedied if OMB required agencies to report all research activity, regardless of budget account, in a consistent manner.

REFERENCES

American Association for the Advancement of Science (AAAS). 2000, 2001, 2002, 2003. AAAS report XXV, XXVI, XXVII, and XXVIII on Research and Development for FY 2001, 2002, 2003, and 2004. Washington, DC: AAAS Intersociety Working Group.

Committee on Water Resources Research (COWRR). 1973 and 1974. Federal Water Resources Research Program for 1973 and 1974. COWRR, FCCSET. Washington, DC: National Science Foundation.

Committee on Water Resources Research (COWRR). 1977. Directions in U.S. Water Research: 1978–1982. Washington, DC: OSTP Federal Coordination Council for Science, Engineering and Technology (FCCSET).

Department of Energy (DOE). 2002. Yucca Mountain Project: Recommendation by the Secretary of Energy Regarding the Suitability of the Yucca Mountain Site for a Repository under the Nuclear Waste Policy Act of 1982. Washington, DC: Office of Civilian Radioactive Waste Management.

Department of Energy (DOE). 2003. The Department of Energy Strategic Plan. DOE/ME–0030. Washington, DC: DOE.

National Research Council (NRC). 1981. Federal Water Resources Research: A Review of the Proposed Five-Year Program Plan. Washington, DC: National Academy Press.

National Research Council (NRC). 2001. Envisioning the Agenda for Water Resources Research in the Twenty–First Century. Washington, DC: National Academy Press.

National Science Foundation. 2002. Long–Term Ecological Research Program: Twenty–Year Review. Available at http://intranet.lternet.edu/archives/documents/reports/20_yr_review/#1.0.

U.S. General Accounting Office (GAO). 2001. Water Infrastructure: Information on Federal and State Financial Assistance. GAO 02–134. Washington, DC: GAO.

U.S. General Accounting Office (GAO). 2002. Water Infrastructure: Information on Financing, Capital Planning, and Privatization. GAO 02–764. Washington, DC: GAO.

U.S. Geological Survey (USGS). 2002. U.S. Geological Survey Strategic Plan FY2000–2005. Washington, DC: Department of the Interior. http://www.usgs.gov/stratplan/stratplan_rev.pdf.

U.S. Geological Survey (USGS). 2004. Mission of the Water Resources Discipline. Washington, DC: Department of the Interior. http://water.usgs.gov/welcome.html.

U.S. Government Printing Office. 2003. The Budget for Fiscal Year 2004, Historical Tables. Washington, DC.

5

Data Collection and Monitoring

The long-term monitoring of hydrologic systems and archiving of the resulting data are activities that are inseparable from the water resources research enterprise of the nation. Data are essential for understanding physicochemical and biological processes and, in most cases, provide the basis for predictive modeling. Examples of water resources and other relevant data that are collected through a variety of measurement devices and networks are:

- hydrologic storages and fluxes such as soil moisture, snow pack depth, precipitation, streamflow, hydraulic head, recharge, and evapotranspiration
- land-ocean-atmosphere energy fluxes
- water, land, and air quality measures, including physical, chemical, biological, and ecological elements
- water and energy demand, consumptive use, and return flows
- terrain elevation and land use, and lake, stream, and river geometry

Data collection is the means by which these types of data are acquired for multiple uses, including for flood warnings and other health and safety monitoring activities, weather prediction, engineering design, commercial and industrial applications, and scientific research. Monitoring is data collection with the more targeted purpose of detecting and drawing attention to changes in selected measures, particularly extreme changes. Monitoring data have multiple applications. They may serve as indicators of health and safety risks, as trip wires for policy changes, or as the basis for research on variability and trends in hydrologic and related phenomena.

The full dimensions of the challenges and opportunities associated with data collection and management for the hydrologic sciences have become evident in recent years. These issues are especially important for federal agencies because these agencies are instrumental in developing new monitoring approaches, in validating their efficacy through field studies, and in managing nationwide monitoring networks over long periods.

This chapter is not a comprehensive assessment of water resources data collection activities. Rather, it is intended to highlight the importance of data collection and its role in stimulating and facilitating water resources research. Thus, it relies upon a few specific examples from certain federal agencies. As a consequence, not all data collection activities relevant to water resources research are included (e.g., active disease surveillance and monitoring of land use are not discussed), nor are all federal and state agencies that support or actively conduct monitoring mentioned.

CHALLENGES IN MONITORING

There are important challenges facing federal agencies that collect and manage hydrologic data. One of these challenges is related intrinsically to the types of problems for which hydrologic data are being used. For example, the analysis of problems related to floods and droughts requires specific information about extreme events, which can be developed only after conducting decades or even centuries of precipitation and streamflow monitoring across a variety of different climatic and hydrologic settings. Similarly, an assessment of the impact of global climate change on groundwater and surface water resources will require basic monitoring systems capable of providing data for time periods of centuries. With other problems, like nonpoint source contamination of streams resulting from runoff laden with nitrate, pesticides, and sediment, hourly data may be required because of the close association of stream contamination with the timing of storms and resulting runoff processes. In this case, the greater challenge is encompassing all relevant spatial scales, because the local variability in contaminant loading is related to changes in land use and farming practices. In general, the broad spectrum of present and future scientific water problems nationwide requires monitoring systems that function reliably over both large and small time and space scales.

Unfortunately, as described in detail in later sections, observational networks to measure various water characteristics have been in decline during the last 30 years because of political and fiscal instabilities (e.g., NRC, 1991; Entekhabi et al., 1999). The following sections provide a detailed discussion of how several national monitoring networks have fared over time. The funding situation for monitoring networks is remarkably similar to that for research on improving data collection activities (Category VII), which Chapter 4 showed as having declined to a level that is only a fourth of its value in the mid 1970s. These facts point to the need for increased funding for *in situ* data networks as a necessary complement to water resources research activities.

Unquestionably, the complexity of monitoring has increased dramatically as researchers have begun to sort out the interactions among physical, chemical, and biological processes in water (Pfirman and AC-ERE, 2003). Before 1960, hydrologic monitoring in the U.S. Geological Survey (USGS) emphasized streamflows, sediment transport, and groundwater levels. As problems of contamination became more evident, monitoring expanded to examine basic water quality variables. Now, monitoring encompasses a variety of anthropogenic compounds in water and sediment, aquatic organisms, and other biological characteristics. In just the past few years, advances in analytical techniques and the discovery of new, ecologically significant families of contaminants have made the need for comprehensive monitoring of aquatic systems even more evident.

Federal agencies are under significant pressure to respond to the increasing need for chemical and biological monitoring of aquatic systems. The creation of the National Water Quality Assessment (NAWQA) program of the USGS highlights the emerging importance of this type of monitoring, as well as its complexity and potential costs. Yet, even this large program is only a first step in providing the information necessary to support federal regulations related to water quality (e.g., the Total Maximum Daily Load program), to support restoration of aquatic ecosystems impacted by agriculture (e.g., the Neuse River basin), and to promote the sustainability of water resources (e.g., the Rio Grande). The development of the kind of enhanced chemical and biological monitoring that will be needed to address such issues remains a challenge.

Complexity in monitoring also arises because of the scales at which problems are now manifested. For example, the study of hypoxia in the northern Gulf of Mexico is inexorably linked to the Mississippi–Atchafalaya River Basin, which drains an area of 3.2 million square miles (NSTC, 2000). New, scale-appropriate techniques will be required to examine hydrologic conditions across watersheds of subcontinental proportions. Unfortunately, such large-scale monitoring approaches are in their infancy and are not sufficiently developed to meet even immediate needs for some classes of problems. As will be discussed shortly, this challenge to provide new kinds of data is also an opportunity for innovative research related to space-borne and sensor technologies.

The increasing variety and quantity of information coming from monitoring systems have created new problems of data warehousing and dissemination. Federal and state agencies over time have developed important databases (e.g., the Environmental Protection Agency's [EPA] STORET and USGS' NWIS) that are an increasingly useful way to identify problems and research needs. However, as a consequence of historical agency responsibilities and/or lack of national funding, federal water resources data are spread among different federal and state agencies with broadly different capabilities for supporting user needs and with different resources for making important legacy data available to researchers. Taylor and

Alley (2001) point out that many state agencies have backlogs of data waiting to be summarized in electronic databases. (Indeed, one agency representative reported that summer students were used to develop web sites that can disseminate valuable historical data from research watersheds.) There is little consistency in federal and state programs for recovering the costs of monitoring. Some monitoring data (e.g., some stream discharge data) are freely available, while in other cases (e.g., Landsat), Congress has mandated cost recovery for the data, which makes using such data expensive in research and may limit their creative use.

An important aspect of improving hydrologic data collection is to take into account the uncertainties associated with both data collection methods and the design of data collection networks. Hydrologic forecasting relies heavily on measurements of multiple hydrologic variables taken over long periods of time, particularly because low-frequency but high-magnitude events can have near-irreversible effects on water supply. As monitoring networks decrease in size and density, uncertainty increases. Methods of rescuing and augmenting the available data are needed to reduce the uncertainty of predictions. Similarly, methods of validating and estimating the uncertainty of remotely sensed data are needed, as these sources of data are becoming increasingly important in hydrologic analyses.

OPPORTUNITIES IN MONITORING

Hydrologists working in the United States can now take advantage of new monitoring data and new technical approaches for monitoring hydrologic processes. Important advances for the diagnosis and prediction of hydrologic processes have come from remote sensing using products derived from observations that span a wide range of the electromagnetic spectrum (e.g., visible, infrared, microwave) (Owe et al., 2001). For example, Landsat satellites have provided a significant record of land-cover conditions on the earth's surface and an ability to monitor land-use changes (e.g., Running et al., 1994), and they can measure water clarity and chlorophyll in lakes (www.water.umn.edu). Remote sensing has also been shown to provide significant information about hydrologic extremes, such as drought, and to have the potential to enhance our ability to forecast these events (e.g., Kogan, 2002). New satellite sensors (e.g., Advanced Microwave Sounding Unit on the National Oceanic and Atmospheric Administration [NOAA] 15, 16, and 17 satellites) with the ability to penetrate cloud cover and produce land surface moisture products with temporal resolutions of hours and spatial resolutions of tens of kilometers promise to enhance our operational database that supports water supply forecasts (Ferraro et al., 2002). Chlorophyll levels in freshwater lakes now are being mapped routinely by satellite. Furthermore, high-resolution satellites are being used to map distributions of different types of aquatic plant communities in wetlands and littoral areas of lakes, and aircraft-mounted spectroradiometers are being used to map aquatic vegetation and water quality conditions (e.g., turbidity, phosphorus, and chlorophyll *a*) in rivers.

Radar altimeters, like TOPEX/Poseidon and Jason-1, which have been used mostly for measuring changes in sea level, are also useful for measuring the stages of large rivers and lakes (Birkett, 1995, 1998). The laser altimeters aboard ICESat, although not particularly reliable at the present stage of development, have demonstrated the potential to measure water-surface elevations for small waterbodies on a regular basis. The Gravity Recovery and Climate Experiment (GRACE) mission provides unprecedented capabilities to assess changes in water storage over large regions (Rodell and Famiglietti, 1999, 2002). GRACE is anticipated to provide monthly measurements of total water storage anomalies with a spatial scale of longer than several hundred kilometers, and with an accuracy of a few millimeters in water-height change.

Ground-based remote sensing monitoring systems have been developed that are applicable to water resources investigations of regional water supplies. A good example is the widespread national deployment of the WSR-88D weather radar (NEXRAD). Each NEXRAD station monitors thousands of square kilometers and provides almost continuous space-time estimates of precipitation with kilometer resolution (e.g., Klazura and Imy, 1993). When properly calibrated, these systems can provide highly resolved estimates of precipitation for complex storms or for regions where coverage by conventional gages is limited (e.g., Seo et al., 2000).

Remote sensing will undoubtedly change the way that some hydrologic monitoring is carried out, although routine use of some of these technologies for water resources research is still years away. In the meantime, the use of the remotely sensed data for water resources research on regional and local scales will require validation and adjustment with legacy monitoring measurements and, for many applications, combined use with *in situ* measurements of longer records. For example, it is expected that detailed monitoring of chemical and biological conditions will still require sampling and laboratory analyses or *in situ* sensor measurements, and likely some combination of both. Fortunately, technological advances are being made in developing low-cost and reliable sensors and miniaturized *in situ* instruments to measure a wide variety of chemical and biological contaminants in natural waters (ASLO, 2003).

In addition to requiring validation with legacy monitoring, profitable use of remotely sensed data will require measurements of associated space-time observation uncertainty. Such efforts are required because remote sensors in many cases do not directly measure the quantity of interest, because the data they generate carry biases due to atmospheric and land surface interference, or because their penetration depth into the land surface is shallow. In addition, regional and local databases of remotely sensed data in most cases have short record lengths that are often inadequate for the study of water supply variability, including climate extremes. It should be noted that remote sensing systems for some crucial water monitoring data—for example biological metrics of water quality like numbers and health of algae, invertebrates, fish, and aquatic and terrestrial plants—are only now being developed.

Thus, although there is much progress with respect to remote sensing systems measurement and monitoring, important *in situ* monitoring systems are in decline without clear plans for transitioning to new or alternative technologies. Because of the "stovepiping" within agencies, there is no overall coordination in operating monitoring systems and in determining directions of new technological initiatives.

A new paradigm for data acquisition and management is offered by the cyberinfrastructure view of information, which is a significant step for enabling the next generation of research in science and engineering. The National Science Foundation (NSF) is at the forefront of this important new initiative, which recognizes the ability of new developments in information technologies to change the way data are collected, managed, and used (Atkins et al., 2003). The term cyberinfrastructure refers to the variety of approaches (some new, some old) for the creation, dissemination, and preservation of knowledge. The NSF vision is "to use cyberinfrastructure to build more ubiquitous, comprehensive digital environments that become interactive and functionally complete for research communities in terms of people, data, information tools, and instruments" (Atkins et al., 2003). The development of these approaches is a bold step forward in the seamless integration of experimentation and data collection.

The most important implication of cyberinfrastructure for water resources is that monitoring is not simply an isolated task but is part of an integrated information strategy that more directly connects researchers with data and the actual process of measurements. If properly executed, this strategy has the potential to create a more uniform technological vision among federal agencies and to reduce redundancies in data handling among federal agencies. In other words, there are opportunities for government agencies concerned with monitoring to develop a systemic approach to handling the explosion of new data and the operation and maintenance of monitoring systems. Federal agencies are relatively independent in their approaches and solutions to issues of monitoring and data management.

STATUS OF KEY MONITORING PROGRAMS

Addressing water resources concerns in the future will require increasingly sophisticated monitoring data. Streamflow data, for example, are necessary to (1) support important public policy decisions concerning towns located in the floodplains of rivers like the Mississippi, (2) engineer structures to limit flood damages from rivers like the Red River of the North, and (3) manage water resources in important western rivers like the Colorado. Chemical, biological, and sediment data are needed to evaluate the efficacy of attempts to restore water quality and ecological health in the Chesapeake Bay, the Mississippi River system, and the San Francisco Bay, which are being adversely affected by nonpoint source contamination. Global climate change is predicted to have major impacts on rivers of the northwestern United States and the Prairie Pothole region of the northern Great Plains (NAST, 2000) and will require new monitoring networks.

The following sections provide examples of the national decline in some types of monitoring systems and the funding limitations that have stifled efforts toward new national systems in groundwater and soil-moisture monitoring. The lack of investment in hydrologic monitoring is hard to reconcile, given the societal cost and significance of problems affecting water resources now and in the future. The reasons for the decline are many and speculative. It is possible that the distributed nature of water resource problems and the long periods between extreme events have obscured the need for monitoring systems (i.e., it is too early to see the negative impacts of the cuts in monitoring). Alternatively, it may be that the consequences of dismantling or substantially reducing monitoring systems have been small to date because accumulated science and engineering knowledge has been able to cope to some extent with the uncertainties of reduced observations. Another possibility that might account for the observed decline is that there are likely fewer and less visible investments in large water resources structures (e.g., dams, reservoirs, canals, and reclamation projects) that might require data.

The responsibilities for collecting, maintaining, and distributing hydrologic data remain with federal agencies, and this situation is not likely to change, given the nature of water resources data collection as a public good (see Chapter 1). The dilemma of the federal agencies is how to simultaneously maintain legacy monitoring systems, respond to escalating needs for expanded monitoring of all kinds, and take advantage of new opportunities in infrastructure development for measurement and data management.

Streamflow

The USGS has been collecting streamflow information since 1889 and today operates a national network of about 7,200 stream gages. The information provided by the network is used for many purposes, including water resource planning, daily water management, flood prediction and hazard estimation, water quality assessment and management, aquatic habitat assessment and mitigation, engineering design, recreation safety, and scientific research. Funding for the network comes from the USGS and over 800 other federal, state, and local agencies. This unique arrangement helps ensure streamflow information relevant to local needs; however, it also means that the USGS does not have complete control over the network, including the number or location of the individual stream gages that constitute the network.

Stream gages with long periods of record are of great importance for estimating hydrologic extremes (floods and droughts) and for resource planning. These gages are crucial to describing and understanding the effects of climate, land-use, and water-use changes on the hydrologic system. However, maintaining stream gages with a long period of record is not always a priority of many partner agencies. As a consequence, there has been an alarming loss of stream gages with 30 or more years of record over the last three decades, even though the total number of stream gages in the network has remained relatively constant (Norris, 2000). In the period 1990–2001, 690 stream gages with 30 or more years of record were discontinued; nearly 170 of those were discontinued in 1995 alone.

The USGS has described the instability of the national stream gaging network as being related to the current dominant funding process of stream gages and has proposed plans to modernize, stabilize, and fill critical gaps in the network (USGS, 1999; Hirsch and Norris, 2001). This plan, called the National Streamflow Information Program (NSIP), would provide for a stable nationwide backbone stream gage network that would be fully funded by the USGS to maintain for future generations the important long-term streamflow information at critical locations. Stream gages required for local needs would supplement this backbone network and would remain funded through the Cooperative Water Program. In addition to providing a stable component to the national stream gaging network, NSIP would also enhance the value of all streamflow information obtained by the USGS by improving and modernizing data delivery, obtaining more information during hydrologic extremes than is currently obtained, analyzing streamflow information to provide insights on key characteristics (e.g., long-term trends and their relationship to natural and anthropogenic features within watersheds), and conducting research and development aimed at improving instruments and methods in order to provide more accurate, more timely, and less expensive streamflow information in the future. According to NRC (2004), the NSIP program adequately takes into account both the spatial distribution (in terms of value and need) of gages and the use of modeling to provide information about ungaged locations. This is an ambitious program that would require a considerable long-term financial commitment from Congress and the executive branch. So far, only minor additional funding has been appropriated for this program.

Groundwater Levels

Water-level measurements from observation wells are the principal source of information about the effects of hydrologic stresses on groundwater systems. In recent years, the USGS and many state and local agencies have experienced difficulties in maintaining long-term water-level monitoring programs because of limitations in funding and human resources. A poll of USGS district offices and 62 state and local water management or regulatory agencies about the design, operation, and history of long-term observation wells in their respective states indicated that there are about 42,000 observation wells in the United States with five years (a relatively short period) or more of water-level record data (Taylor and Alley, 2001). About a quarter of those are monitored under the USGS Cooperative Water Program. The level of effort in collecting long-term water-level data varies greatly from state to state, and many of the long-term monitoring wells are clustered in certain areas. Although difficult to track, the number of long-term observation wells appears to be declining. For example, the number of long-term observation wells monitored by USGS has declined by about half from the 1980s to today (Alley and Taylor, 2001).

Groundwater-level data become increasingly valuable with length and continuity of record. Ease of access to the data and their timeliness also are valuable, especially during periods of stress such as droughts. Although real-time surface water data have been available through the Internet for nearly a decade, the availability of real-time groundwater data is relatively new within the USGS. Between the years 2000 and 2003, real-time data for wells available through the Internet went from fewer than 300 wells (mostly in south Florida) to nearly 700 wells. Real-time data applications allow effective aquifer management, produce high-quality data, and can be cost-effective. As the availability and reliability of real-time groundwater data increase, so will their value to scientists and the public.

Despite the existence of thousands of observation wells across the nation, most groundwater-level data collection is funded to address state and local issues. Yet, there is evidence that more groundwater problems are becoming regional or national in scale, as exemplified by interstate conflicts over groundwater sources becoming salinized or overdrawn. A case in point is the High Plains aquifer, which encompasses eight states in the central United States. In parts of Kansas, New Mexico, Oklahoma, and Texas, current groundwater withdrawals, primarily used for irrigation, are unsustainable because the natural recharge is low, resulting in a dramatic decline in groundwater levels (Alley et al., 1999). Given the increasing reliance on groundwater sources (Glennon, 2002), it is important to understand trends in groundwater levels and quality over large regions. Unfortunately, there is no comprehensive national groundwater-level network with uniform coverage of major aquifers, climate zones, or land uses. In fact, data on groundwater levels and rates of change are "not adequate for national reporting" according to the report *The State of the Nation's Ecosystems* (H. John Heinz III Center, 2002). Data are not collected using standardized approaches at similar spatial or temporal scales, and the long-term viability of the data collection efforts is uncertain. Ideally, a comprehensive groundwater-level network is needed to assess groundwater-level changes, the data from which should be easily accessible in real time.

Soil Moisture

It has been long recognized that soil moisture in the first one or two meters below the ground surface regulates land-surface energy and moisture exchanges with the atmosphere and plays a key role in flood and drought genesis and maintenance (e.g., Huang et al., 1996; Eastman et al., 1998). Soil moisture deficit partially regulates plant transpiration and, consequently, constitutes a diagnostic variable for irrigation design (e.g., Dagan and Bresler, 1988). High extremes of soil moisture are associated with high potential for flooding and hazardous conditions. As the "state variable" of the vadose zone, soil moisture plays a key role in surface–subsurface water exchanges. Although the importance of soil moisture for hydrologic science and applications cannot be overemphasized, there are few long-term and large-scale measurement programs for soil moisture that provide *in situ* profile data suitable for hydroclimatic analysis and design in the United States (e.g., Hollinger and Isard, 1994; Georgakakos and Baumer, 1996) and abroad (e.g., Vinnikov and Yeserkepova, 1991). Active and passive microwave data from polar orbiting satellites or reconnaissance airplanes do provide estimates of surface soil moisture with continuous spatial coverage. However, they are limited in that they only measure soil moisture within the first few centimeters of the soil surface, and they are reliable only when vegetation cover is sparse or absent (e.g., Ulaby et al., 1996; Jackson and Le Vine, 1996).

This lack of long-term soil moisture data over vast areas of the United States affects how well soil moisture is incorporated into hydrologic models for watersheds or large regions. At the present time, models must use estimates derived from secondary sources of information or from other models, rendering predictions pertaining to ecosystem behavior or surface water–groundwater interactions subject to significant uncertainty. Even a few long-term monitoring networks of soil moisture would substantially decrease the uncertainty in predicting processes that critically depend on soil moisture levels (like flow, water chemistry, and plant response). In a similar vein, the uncertainty of predictive models for managing water supply in western streams reflects the density of streamflow and rainfall monitoring networks, because the amount and the quality of data in areas characterized by high spatial variability in precipitation determine how reliable and precise such models can be.

The development of a national soil moisture monitoring network is an essential element to conducting successful research on the physical, chemical, and biological processes in the surface layer of the continental United States. The U.S. Department of Agriculture (USDA) Natural Resources Conservation Service has established a coordinated national network of *in situ* measurements of soil moisture and soil temperature in support of agricultural needs (Soil Climate Analysis Network or SCAN). Although this is a step in the right direction, significant expansion of the network into nonagricultural areas together with a long-term commitment for high quality data are necessary for water resources analysis on climatic and regional scales. Furthermore, modeling and observational studies have shown substantial soil moisture variability over a range of scales (e.g., Rodriguez-Iturbe et al., 1995; Guetter and Georgakakos, 1996; Vinnikov et al., 1996; Lenters et al., 2000), and the development of a monitoring plan for soil moisture on the basis of both remotely sensed and on-site data is a requisite research endeavor that should account for such variability.

Water Quality

Water quality monitoring has seen declining trends in funding similar to those for streamflow and groundwater level monitoring. Both the EPA, through its Environmental Monitoring and Assessment Program (EMAP), and the states, through their delegated authority from EPA for the Clean Water Act, conduct some water quality monitoring. Unfortunately, efforts are inconsistent from state to state and are often inadequate, and some are in decline. Recent reports (e.g., GAO, 2000; NRC, 2002a; Mehan, 2004) cite the need for consistent ambient water quality data for purposes of Clean Water Act compliance. The current shortfalls and future needs are illustrated below using USGS water quality programs as examples.

For some areas, USGS data are relied upon to establish trends in water quality over time and to compare conditions across local jurisdictional boundaries. Unfortunately, water quality monitoring networks within USGS have not received additional funding since their inception and thus have been declining simply due to the impacts of inflation. USGS surface water quality networks include the Hydrologic Benchmark Network (HBN), operating since 1964, and the National Stream Quality Accounting Network (NASQAN), operating since 1973. HBN monitors small watersheds in areas relatively free from human impacts, from water diversions, and from water impoundments, providing important baseline data for understanding water quality impairments and needed improvements. When it first began in 1964, HBN water quality sampling was quarterly to monthly depending on location, but since 1997 there has been no water quality sampling at the 52 watersheds (Mast and Turk, 1999). Beginning in 2003, limited sampling resumed at 15 of the remaining 36 watersheds.[1]

NASQAN measures water quality in the nation's largest river systems, and it also includes many coastal drainages. At the program's operational peak (in 1976), more than 500 locations were sampled either monthly or six times a year for major ions, nutrients, trace metals, indicator bacteria, and periphyton. Now, only 33 sites remain, with sampling at a frequency adequate for annual flux estimates and with new capabilities for sampling some pesticides. These networks now provide a fraction of the data they once produced (though some prior components may be handled by the NAWQA program—see below).

Assessing water quality, both for trends over time and for causative factors, is growing in importance for water resource management. The NAWQA program provides data and information on the most important (defined by water use for municipal supply or irrigated agriculture) river basins and aquifer systems (study units). NAWQA comprehensively samples surface water and groundwater for physical and chemical variables, and it produces data on aquatic communities of

[1]The program was reduced from 52 watersheds to 36 because USGS appropriations have been level or declining for several years, while costs increase about 4 percent every year (Robert Hirsch, Chief Hydrologist, USGS, personal communication, 2004).

fish, insects, and algae. Originally NAWQA was designed for sampling 59 study units, but by 2001 only 51 were sampled, and now because of funding constraints only 42 study units continue to operate.

Within NAWQA, many sampling activities have been curtailed. For example, between 1993 and 2001, more than 600 fixed-station surface water sites were sampled, but now only about 150 sites continue to operate. These sites provide the only continuous water quality trend sampling. Although thousands of bed sediment and tissue data were collected from streams in the first years of sampling, most of those sampling efforts cannot be repeated. Groundwater sampling between 1993 and 2001 included more than 6,500 wells. There are not adequate resources to sample those sites repeatedly for trends, and only 2,400 wells are being resampled in urban, agriculture, and large aquifer networks, reducing substantially the density of the sampling network. Although the numbers of wells may seem large, the trend network provides insight at an average of less than 60 wells per study unit (the median study unit area is 21,000 sq. mi.). The recent National Research Council (NRC) report (NRC, 2002a) states that NAWQA cannot decrease its number of study units further and still provide the national scope of data called for by Congress.

Monitoring activities related to sediment are particularly crucial because it is the most widespread pollutant in U.S. rivers (EPA, 2002). The USGS is the principal source of fluvial-sediment data, providing daily suspended-sediment discharge data at about 105 sites (sediment stations) in 2002. These data serve traditional uses that include design and management of reservoirs and in-stream hydraulic structures and dredging. In the last two decades, information needs have expanded to include those related to contaminated sediment management, dam decommissioning and removal, environmental quality, stream restoration, geomorphic classification and assessments, physical–biotic interactions, the global carbon budget, and regulatory requirements of the Clean Water Act including the EPA's Total Maximum Daily Load program.

An increasing need for fluvial-sediment data has coincided with a two-thirds decline in the number of USGS sediment stations from the peak of 360 in 1982 (Gray, 2002) to about 105 sites now. Among the factors cited for the decline in the number of USGS daily sediment stations was the need for less expensive and more accurate fluvial-sediment data collected using safer, less manually intensive techniques. Any decrease in sediment monitoring should be of particular concern given that the physical, chemical, and biological sediment damages in North America alone were estimated at $16 billion in 1998 (Osterkamp et al., 1998). In its review of the NAWQA program, the NRC noted the serious need to improve sediment monitoring (NRC, 2002a).

Box 5-1 discusses how the lack of reliable water quality monitoring data has hampered efforts to sensibly plan for development in New Jersey and comply with state laws. It exemplifies not only shortages in groundwater quality data, but also in flow data.

BOX 5-1
Water Quality Monitoring in the Pinelands National Reserve

In 2000, the New Jersey state legislature passed legislation authorizing the expenditure of $5 million for a study of the impacts of potential groundwater withdrawals on the ecology of the Pinelands National Reserve. The Pinelands National Reserve, an area of about 1 million acres, occupies the southern third of the state and is underlain by the Cohansey, an extensive groundwater aquifer. Development pressure in the lands surrounding the preserve, including the suburbs of Philadelphia, Pennsylvania, and Camden, New Jersey, the burgeoning Cape May peninsula, and the Atlantic City region, are intense. Building and economic activity in this area are increasingly limited by water availability; the aquifers currently being utilized are already overpumped and salinized. The Pinelands Reserve is protected by both state and federal legislation that specifies the protection of "the natural ecological character of the region" as the criterion for setting land-use policy. This large-scale (multistate) study was authorized in order to determine whether exports of water from the Pinelands watersheds to the surrounding developing areas would negatively affect the aquatic ecosystems of the Pinelands.

During initial discussions about the scope of the research, the scientists involved (from the state Pinelands Commission, the New Jersey Division of the U.S. Geological Survey, and Rutgers University) decided that it was important to study watersheds across the range of aquifer and land-use conditions in the Pinelands. Thirty-five (35) subbasins were initially identified as potential intensive study areas. They represented the range of aquifer thicknesses (from less than 100 feet to over 500 feet), current rates of pumping (from 0 to 740 megagallons per year), land uses (from nearly complete forest cover to mixtures of developed and agricultural land with little forest cover), amounts and types of wetland, and stream lengths. Despite the large range of existing conditions in the subbasins, all but two of the subbasins were excluded from consideration as intensive study sites because they lacked the long-term monitoring data of both groundwater flow and water quality necessary to calibrate hydrologic, chemical, and ecosystem water balance models. Two other subbasins had partial records from continuous-flow monitors, and 15 subbasins had partial, discontinuous records from low-flow gages. Only nine subbasins had water quality monitoring data, and for only one of these was there a long-term continuous record.

One of the subbasins with both long-term continuous flow and water quality records, McDonald's Branch, is situated in the center of the region, is small and forested, has variable aquifer thickness, and lacks some of the wetland types of importance to the study. The other subbasin with

continued

BOX 5-1 Continued

long-term continuous water flow and water quality data, the East Branch Bass River, is much larger, with some urbanized land within the basin; however, it lacks some of the major aquatic communities that are the focus of the study. Thus, the lack of continuous water flow and water quality monitoring data from all but two of the subbasins prevented the research team from studying the range of conditions that would best answer the critical management questions posed by the legislation. Moreover, the team was constrained to use two subbasins that are not strictly comparable, in that each lacks an important component of the hydroecological system that is the target of this large-scale investigation.

Water Use

Estimating the demands for water and amounts withdrawn from various surface-water and groundwater sources is of critical importance to water resources management. Since 1950, the USGS has compiled and disseminated estimates of water use for the nation at five-year intervals. Most of the data are collected, however, not by the USGS but by the individual states to support their water-use permitting and registration programs. Although matching funds for the analysis and aggregation of water use data are often available through the USGS Cooperative Water Program, some states make little effort in this area. Thus, the quality of water use data varies considerably from state to state. Unfortunately, because of funding limitations, the USGS had to reduce the scope of reporting on a nationwide basis in 2000 for several categories of water use. Reductions in national scope included (1) eliminating estimates of commercial use and hydroelectric power (which is counter to recommendations in the *Envisioning* report [NRC, 2001]), (2) providing information for mining, livestock, and aquaculture only for large-use states, (3) eliminating estimates of consumptive use and public deliveries, and (4) compilation at the county, but not watershed, level.

In a recently completed NRC review of the USGS National Water Use Information Program (NWUIP) (NRC, 2002b), basic questions about the nature of water use, the water use information needed in the United States, and the USGS role in generating and disseminating that information were considered. Major recommendations contained in the NRC report include the following:

- The NWUIP should be elevated to a *water use science* program (rather than a water use accounting program), emphasizing applied research and techniques development in the statistical estimation of water use and the determinants and impacts of water-using behavior.

• To better support water use science, the USGS should build on existing data collection efforts to systematically integrate datasets, including those maintained by other federal and state agencies, into datasets already maintained by the NWUIP.

• The USGS should systematically compare water use estimation methods to identify the techniques best suited to the requirements and limitations of the NWUIP. One goal of this comparison should be to determine the standard error for every water use estimate.

• The USGS should focus on the scientific integration of water use, water flow, and water quality to expand knowledge and generate policy-relevant information about human impacts on both water and ecological resources.

• The USGS should seek support from Congress for dedicated funding of a national component of the recommended water use science program to supplement the existing funding in the Cooperative Water Program.

* * *

The preceding section discussed some of the important data collection networks relevant to water resources research, but the discussion was not intended to be exhaustive. For example, both public health data collection (e.g., active disease surveillance) and monitoring of climate variables like precipitation were not discussed. This should not be interpreted as implying that they are less important types of data. Indeed, Box 5-2 discusses how federally funded monitoring for climate indices is routinely used to manage water resources in the Florida Everglades.

CONCLUSIONS AND RECOMMENDATIONS

Monitoring efforts are inseparable from the research efforts described in other chapters of this report. Furthermore, they are critical to addressing water resources problems related to floods and droughts, agricultural sustainability, global climate change, and other high priorities for water resources research (as expressed in Chapter 3). Indeed, the continuing need for high-quality, long-term *in situ* data was the only water resources-related issue expressed unanimously by 13 state government representatives who addressed the committee in January 2003 (see Appendix D).

BOX 5-2
Real-Time Water Management in South Florida

The South Florida Water Management District (District) is one of the largest nonfederal water management agencies in the country with far-reaching authority over water use and environmental protection from Orlando to Key West. The area is blanketed by a mammoth series of federal water projects constructed by the U.S. Army Corps of Engineers primarily in the 1950s and 1960s. In recent years the District has developed unique approaches to operating federal facilities to achieve environmental benefits that were not valued when the project was first built, but without compromising the water supply and flood control requirements of the original project.

Lake Okeechobee is at the heart of the water management system in South Florida, storing floodwater from the upstream watershed and supplying water for agriculture, urban populations, and the Everglades. It is also an indispensable natural resource that contains a contiguous 90,000-acre wetland system supporting thousands of wading birds and numerous endangered species, as well as supporting the sport fishing that is critical to the local economy. The need to manage this resource for competing and sometimes conflicting objectives has led the District to consider the results of the latest federally supported climate research in the process of making operational decisions. Both seasonal and multiseasonal climate outlooks produced monthly by the Climate Prediction Center of the National Weather Service have been incorporated into operations of the regional water management project. The District's approach links local and global climate indices to on-the-ground hydrologic information to make weekly adjustments in water control for Lake Okeechobee. Many institutions around the country have active research programs, but few have been successful in implementing climate forecast information into day-to-day operations. Further, the District has taken the important step of revising the official water control manuals to formalize the routine employment of information provided by climate research programs associated with the National Oceanic and Atmospheric Administration and others. Without assistance from several federal research entities, the successful implementation of climate-based operations in south Florida, and the public's acceptance of using innovative operational planning methods that employ forecasts, would not have been possible.

The challenges to the monitoring of water resources are formidable and include a sizable increase in the number of features to be monitored, as the variety, scope, and complexity of problems expand, especially with respect to biological issues; a major expansion in the range of time/space scales that must be addressed by data collection and monitoring as new, major problems develop; and difficulties in making the increasingly large amounts of data available quickly and efficiently. The following conclusions and recommendations address the need to match these challenges, as well as emerging water resources problems, with new investments for basic data collection and monitoring.

Key legacy monitoring systems in areas of streamflow, groundwater, sediment transport, water quality, and water use have been in substantial decline and in some cases have been nearly eliminated. These systems provide data necessary for both research (i.e., advancing fundamental knowledge) and practical applications (e.g., for designing the infrastructure required to cope with hydrologic extremes). Despite repeated calls for protecting and expanding monitoring systems relevant to water resources, these trends continue for a variety of reasons.

The consequences of the present policy of neglect associated with water resources monitoring will not necessarily remain small. New hydrologic problems are emerging that are of continental or near continental proportions. The most obvious are the likely impact of global climate change on water resources; hypoxia in the Gulf of Mexico, related to nutrient loading from the Mississippi River; and the questionable sustainability of the water supply for western and southern regions given population increases and recent interest in restoring aquatic ecosystems (e.g., the Florida Everglades and the Mississippi River basin). The scale and the complexity of these problems are the main arguments for improvements to the *in situ* data collection networks for surface waters and groundwater and for water demand by sector. It is reasonable to expect that improving the availability of data, as well as improving the types and quality of data collected, should reduce the costs for many water resources projects.

Notwithstanding the overall decline in legacy monitoring systems, there are some positive developments that bear on hydrologic monitoring. For example, the NEXRAD system provides unprecedented spatial resolution of rainfall distribution. Efforts have continued to support environmental earth-surface observations with new generations of satellites. Other NASA research missions (e.g., IceSat, TOPEX/Poseidon, GRACE) give positive early indications of their potential as monitoring tools for hydrologic systems. Although these new satellite-based measuring systems have important applications to hydrologic research, they are not yet ready to replace legacy monitoring systems. Moreover, for chemical, biological, and groundwater monitoring, in particular, new *in situ* and remote sensing technologies capable of replacing wells or field sample collections are still in development.

Increases in strategic investments for monitoring are necessary to avoid or at least reduce costs attendant with future water resource or health crises. Investments are required but are not by themselves sufficient to ensure that the data necessary to attack 21st century problems will be available to researchers and policy makers alike. Federal agencies need to adopt a research perspective toward monitoring and data collection that better integrates monitoring with the research efforts described in other chapters of this report. There is also a strong need for cooperation among agencies concerned with collecting, storing, and managing hydrologic data, particularly from research watersheds and legacy monitoring systems. The NSF cyberinfrastructure initiative is an example of a visionary approach for creating comprehensive digital environments linking people and data. It is recommended that an interagency task force concerned with information technology and data management be established and that it develop a non-NSF federal cyberinfrastructure community.

REFERENCES

Alley, W. M., T. E. Reilly, and O. L. Franke. 1999. Sustainability of ground-water resources. U.S. Geological Survey Circular 1186.

Alley, W. M., and C. J. Taylor. 2001. The value of long-term ground water level monitoring. Ground Water 39:801.

American Society of Limnology and Oceanography (ASLO). 2003. Emerging Research Issues for Limnology: the Study of Inland Waters. Waco, TX: ASLO.

Atkins, D. E., K. K. Droegemeir, S. I. Feldman, H. Garcia-Molina, M. L. Klein, D. G. Messerschmitt, P. Messina, J. P. Ostriker, and M. H. Wright. 2003. Revolutionizing Science and Engineering through Cyberinfrastructure. Washington, DC: National Science Foundation.

Birkett, C. M. 1995. The contribution of TOPEX/POSEIDON to the global monitoring of climatically sensitive lakes. JGR–Oceans 100 (C12):25,179–25,204.

Birkett, C. M. 1998. Contribution of the TOPEX NASA radar altimeter to the global monitoring of large rivers and lakes. Water Resources Research 34(5):1223–1239.

Dagan, G., and E. Bresler. 1988. Variability of an irrigated crop and its causes: 3—numerical simulation and field results. Water Resources Research 24(3):395–401.

Eastman, J. L., R. A. Pielke, and D. J. McDonald. 1998. Calibration of soil moisture for large-eddy simulations over the FIFE area. J. Atmospheric Sciences 55:1–10.

Entekhabi, D., G. R. Asrar, A. K. Betts, K. J. Beven, R. L. Bras, C. J. Duffy, T. Dunne, R. D. Koster, D. P. Lettenmaier, D. B. McLaughlin, W. J. Shuttleworth, M. T. van Genuchten, M. Y. Wei, and E. F. Wood. 1999. An agenda for land surface hydrology research and a call for the second international hydrological decade. Bull. Amer. Meteor. Soc. 80:2043–2058.

Environmental Protection Agency (EPA). 2002. National Water Quality Inventory—2000 report: EPA–841–R–02–001. Washington, DC: EPA.

Ferraro, R., F. Weng, N. Grody, I. Guch, C. Dean, C. Kongoli, H. Meng, P. Pellegrino, and L. Zhao. 2002. NOAA satellite–derived hydrological products prove their worth. EOS 83(29):429–438.

General Accounting Office (GAO). 2000. Key EPA and State Decisions Limited by Inconsistent and Incomplete Data. General Accounting Office RCED–00–54. 73 p. Washington, DC: GAO.

Georgakakos, K. P., and O. W. Baumer. 1996. Measurement and utilization of on-site soil moisture data. Journal of Hydrology 184:131–152.

Glennon, R. 2002. Water Follies: Groundwater Pumping and the Fate of America's Fresh Waters. Washington, DC: Island Press.

Gray, J. R. 2002. The need for sediment surrogate technologies to monitor fluvial-sediment transport. Proceedings of the Turbidity and Other Sediment Surrogates Workshop, April 30–May 2, 2002, Reno, Nevada. http://water.usgs.gov/osw/techniques/TSS/listofabstracts.htm.

Guetter, A. K., and K. P. Georgakakos. 1996. Large-scale properties of simulated soil water variability. J. Geophysical Research—Atmospheres 101(D3):7175–7183.

H. John Heinz III Center for Science, Economics and the Environment. 2002. The State of the Nation's Ecosystems: Measuring the Lands, Waters, and Living Resources of the United States. Cambridge, UK: Cambridge University Press.

Hirsch, R. M., and J. M. Norris. 2001. National Streamflow Information Program: Implementation Plan and Progress Report. USGS Fact Sheet FS–048–01.

Hollinger, S. E., and S. A. Isard. 1994. A soil moisture climatology of Illinois. J. Climate 7(5):822–833.

Huang, J., H. M. Van den Dool, and K. P. Georgakakos. 1996. Analysis of model-calculated soil moisture over the United States (1931–1993) and applications to long-range temperature forecasts. J. Climate 9(6):1350–1362.

Jackson, T. J., and D. E. Le Vine. 1996. Mapping surface soil moisture using an aircraft-based passive microwave instrument: algorithm and example. J. Hydrology 184:85–99.

Klazura, G. E., and D. A. Imy. 1993. A description of the initial set of analysis products available from the NEXRAD WSR–88D System. Bulletin of the American Meteorological Society 74:1293–1312.

Kogan, F. 2002. World droughts in the new millennium from AVHRR-based vegetation health indices. EOS 83(48):557–563.

Lenters, J. D., M. T. Coe, and J. A. Foley. 2000. Surface water balance of the continental United States, 1963–1995: regional evaluation of a terrestrial biosphere model and the NCEP/NCAR reanalysis. J. Geophysical Research—Atmospheres 105(D17).

Mast, M. A., and J. T. Turk. 1999. Environmental Characteristics and Water Quality of Hydrologic-Benchmark Network Stations in the Eastern United States 1963–95. U.S. Geological Survey Circular 1173–A. 158 p.

Mehan, T. 2004. Better monitoring for better water management. Water Environmental Research 76(1): 3–4.

National Assessment Synthesis Team (NAST). 2000. Climate change impacts on the United States: The potential consequences of climate variability and change. Washington, DC: U.S. Global Change Research Program.

National Research Council (NRC). 1991. Opportunities in the Hydrologic Sciences. Washington, DC: National Academy Press.

National Research Council (NRC). 2001. Envisioning the Agenda for Water Resources Research in the Twenty-first Century. Washington, DC: National Academy Press.

National Research Council (NRC). 2002a. Opportunities to Improve the U.S. Geological Survey National Water Quality Assessment Program. Washington, DC: National Academy Press.

National Research Council (NRC). 2002b. Estimating Water Use in the United States—A New Paradigm for the National Water-Use Information Program. Washington, DC: National Academy Press.

National Research Council (NRC). 2004. Assessing the National Streamflow Information Program. Washington, DC: The National Academies Press.

National Science and Technology Council (NSTC). 2000. 2000 Annual Report. Washington, DC: NSTC.

Norris, J. M. 2000. The value of long-term streamflow records. Water Resources Impact 2(4):11.

Osterkamp, W. R., P. Heilman, and L. J. Lane. 1998. Economic considerations of a continental sediment-monitoring program. International Journal of Sediment Research 13(4):12–24.

Owe, M., K. Brubaker, J. Ritchie, and A. Rango (eds.). 2001. Remote Sensing and Hydrology 2000. IAHS Publication No. 267. Wallingford, UK: IAHS Press.

Pfirman, S., and the AC–ERE. 2003. Complex Environmental Systems: Synthesis for Earth, Life, and Society in the 21st Century: A Report Summarizing a 10–Year Outlook in Environmental Research and Education for the National Science Foundation. Washington, DC: National Science Foundation.

Rodell, M., and J. Familglietti. 1999. Detectability of variations in continental water storage from satellite observations of the time dependent gravity field. Water Resources Research 35:2705–2723.

Rodell, M., and J. Familglietti. 2002. The potential for satellite-based monitoring of groundwater storage changes using GRACE: the High Plains aquifer, Central U.S. J. Hydrol. 263(1–4):245–256.

Rodriguez-Iturbe, I., G. K. Vogel, R. Rigon, D. Entekhabi, and A. Rinaldo. 1995. On the spatial organization of soil moisture. Geophysical Research Letters 22(20):2757–2760.

Running, S. W., T. R. Loveland, and L. L. Pierce. 1994. A vegetation classification logic based on remote sensing for use in global biogeochemical models. Ambio 23:77–81.

Seo, D.-J., J. Breidenbach, R. Fulton, D. Miller, and T. O'Bannon. 2000. Real-time adjustment of range-dependent biases in WSR-88D rainfall estimates due to nonuniform vertical profile of reflectivity. Journal of Hydrometeorology 1(3):222–240.

Taylor, C. J., and W. M. Alley. 2001. Ground-Water-Level Monitoring and the Importance of Long-Term Water-Level Data. U.S. Geological Survey Circular 1217.

Ulaby, F. T., P. C. Dubois, and J. van Zyl. 1996. Radar mapping of surface soil moisture. J. Hydrology 184:57–84.

U.S. Geological Survey (USGS). 1999. Streamflow Information for the Next Century: A Plan for the National Streamflow Information Program of the U.S. Geological Survey. USGS Open-File Report 99–456.

Vinnikov, K. Y., and I. B. Yeserkepova. 1991. Soil moisture, empirical data and model results. J. Climate 4:66–79.

Vinnikov, K. Y., A. Robock, N. A. Speranskaya, and C. A. Schlosser. 1996. Scales of temporal and spatial variability of midlatitude soil moisture. J. Geophysical Research—Atmospheres 101(D3):7163–7174.

6

Coordination of Water Resources Research

The provision of adequate supplies of clean water is not only a basic need. It is a matter of national security and the underpinning of the nation's economy as well as its ecological functioning. The strategic challenge for the future is to ensure adequate quantity and quality of water to meet human and ecological needs. The growing competition among domestic, industrial–commercial, agricultural, and environmental needs is approaching water gridlock in many areas. Research is a key component to effectively and efficiently addressing the water resources problems that are quickly becoming today's headlines.

The multiple and looming water crises in virtually every region of the nation suggest that the approximately $700 million currently spent on water resources research is not sufficiently focused or is not effectively addressing national needs. Without a clearer national water strategy, there is no adequate way to entirely address this issue. Beyond the total investment, there are both topical and operational gaps in the current water resources research portfolio. Although the federal agencies appear to be performing well on their mission-driven research, most of this work focuses on short-term problems, with a limited outlook for crosscutting issues, for longer-term problems, and for the more basic research that often portends future solutions. As a result, it is not clear that the sum of individual agency priorities adds up to a truly comprehensive list of national needs and priorities.

The many state agencies that the committee heard from (see Appendix D) made clear that while some water issues are local, others are becoming increasing common across the country or are of such a scale that the individual states are not equipped to address them. That is, not all problems are local, even though they might appear to be. This misperception has proved to be a significant barrier to coordinating the federal water resources research enterprise. Furthermore, many of today's most pressing (and expensive) problems, and particularly tomorrow's problems, require broader perspectives because they often go beyond the ability and authority of any one federal agency, both in their scale/size and scope.

This chapter summarizes those factors that encourage or discourage effective coordination of large-scale research programs, the roles and benefits of coordination, and the recent history of coordinating water resources research. It concludes with a description of three possible options to achieve coordination.

ENCOURAGEMENTS/DISCOURAGEMENTS OF COORDINATION

During its third meeting, the committee heard from a panel of representatives associated with coordinating large research programs, including programs for highway research through the Transportation Research Board of the National Research Council (NRC), the U.S. Department of Agriculture's Agricultural Research Service, the National Earthquake Hazard Reduction Program (NEHRP), and the U.S. Global Change Research Program (USGCRP). The panelists were asked to comment on which factors or conditions encourage research coordination and which inhibit it in order to help shed light on an effective model for coordination of water resources research.

Several factors stood out as virtually imperative to successful research coordination. First, a strong sense of the relevance of the research, particularly to decision makers like Congress, is important. Much of the success of the USGCRP was attributed to this factor. A second factor is the availability of sufficient resources to implement coordination. In the case of the Transportation Research Board, stakeholders themselves contribute funds that allow for a coordination mechanism—a circumstance that would be difficult or impossible to reproduce in the water community. A third important facet is a clear legal mandate with broad congressional support, such as the mandates of NEHRP and USGCRP. For example, the NEHRP representative noted that the devastating earthquakes in China in 1975 (Haicheng) and 1976 (Tangshan) contributed to support for U.S. legislation authorizing NEHRP in 1977. The National Earthquake Protection Act mandated tasks for four agencies and a two- to three-year reauthorization cycle.

Other facilitating conditions noted by the panel included having research agendas based on scientific objectives and related to agency missions and mandates—obviously a challenge for research areas like water resources that involve multiple federal agencies. Furthermore, strong leadership despite political changes was cited as important. Several administrative factors were cited, including having the participating agencies play complementary roles, engaging external review panels, and making the agenda-setting process transparent. One panelist felt that placing a coordination committee in an agency that could foster scientific exchanges and professional networks between committee meetings was most effective.

Additional factors the panelists said encouraged coordination included specific tasks and deadlines, which, for example, characterized the multiagency Gulf of Mexico hypoxia task force (see NSTC, 2003, and P.L. 105 383, section 640 (b), for example). Mechanisms for ensuring accountability of the coordination process are also important, for example by documenting successes. One panelist pointed out that interagency communication and wide dissemination of research results were important to help ensure stakeholders that they were benefiting directly from their support for research.

The panelists also commented on those factors that discourage coordination of research. A primary deterrent is that fact that different agencies utilize differing budgeting processes (see Box 6-1). Indeed, even with programs like NEHRP, the participating federal agencies have separate authorizing or allocation committees and separate Office of Management and Budget (OMB) examiners. Another frequently mentioned factor is that participants in coordination meetings often do not have decision-making power, and higher-level individuals either are not interested or are not available. A third hindrance, suggested by anecdotal information, is strong agency territoriality regarding particular research topics. Other factors that were identified as working against coordination included the inability to move money across allocation categories (see Box 6-1), lack of sufficient resources and/or staff, and vague planning for coordination. Given that coordination requires additional staff time and funds that could otherwise be used for research or other activities, the benefits of coordination need to be obvious and substantial.

None of the panelists characterized their own programs as having an ideal coordination mechanism. Some of the challenges mentioned included the importance of paying attention to both long-term and emerging issues, and filling gaps when research needs fall between agency missions. One panelist indicated that addressing a gap is more difficult than reducing duplication, that latter often being addressed in response to stakeholders' mistrust of government's management abilities. Another panelist said that agencies have different definitions of "research" that need to be explicitly stated and clarified. For example, some agencies include technology transfer and training in their concept of research while others do not. Coordination itself can be interpreted differently, with some groups perceiving it as only occurring within the federal government, while others include externally conducted research as well.

Further challenges mentioned by the panelists included initiating and maintaining effective relationships with stakeholders and identifying societal issues. Concerns were raised about the difficulties of understanding what data exist, linking databases from separately designed and operated data collection programs, and managing large datasets.

It is important to note that several impediments to coordination of water resources research are institutional arrangements unlikely to be changed by a recommendation from this committee. Examples include the structure of OMB, the appropriation/allocation structure, and federal accounting and fiscal control

BOX 6-1
Budget Issues that Deter Coordination of Research Across the Federal Agencies

Agency research budgets are assembled each year by starting with a "base" defined as those elements that change only marginally from year to year, and then adding "above base" initiatives that may or may not be supported at the departmental or OMB level. Agencies carefully guard their base. Hence, in the context of interagency coordination, an agency would be unlikely to willingly give up a portion of its base to another agency, even if that other agency would apply the funds to a higher national priority. Such trade-offs are expected to occur at the OMB program director's level, but integration across agencies in OMB, which is largely built along the structure of the departments and agencies, can be difficult.

Further, the structure of the congressional appropriations process discourages the shifting of money between agencies that are funded through different spending bills. Each appropriation bill comes out of its own subcommittee on the House Appropriations Committee and out of a corresponding subcommittee on the Senate Appropriations Committee. The agencies funding most of the research on water resources span multiple appropriations subcommittees, as shown in Table 6-1. As a consequence of this fragmentation of water resources research funding, the reality of the appropriations process is that new research directions would most likely have to be funded with new money rather than out of base appropriations. Practically speaking, there is no fungible pot of money representing water resources research funding for all agencies.

TABLE 6-1 Subcommittee Jurisdiction of the House and Senate Appropriations Committee Responsible for Each Federal Agency Doing Water Resources Research

Agency	Appropriations Subcommittee
Army Corps of Engineers	Energy and Water
Environmental Protection Agency	Housing and Urban Development (HUD) and Independent Agencies
Bureau of Reclamation	Energy and Water; Interior
Department of Energy (civilian)	Energy and Water
National Aeronautics and Space Administration	HUD and Independent Agencies
National Science Foundation	HUD and Independent Agencies
National Oceanic and Atmospheric Administration	State, Justice, Commerce
U.S. Department of Agriculture	Agriculture
U.S. Geological Survey	Interior

requirements. These realities were kept in mind as the coordination models and methods described later in this chapter were developed.

Finally, it is clear that there must be sufficiently strong incentives for agencies to consider coordinating their research agendas. The dominant incentive may be cost savings, time savings, insights and/or knowledge gained by leveraging assets with other agencies, or anticipation of meeting mandated or other high-priority needs that would be difficult to meet without a partnership. Whatever the driving force is, it must clearly outweigh the many potential disadvantages. It is easy to understand why it may take a national or natural crisis to effect change for a task as complex and challenging as coordinating water research to meet long-term national needs.

PURPOSES OF COORDINATION

From a federal perspective, the most compelling need for coordination among agencies conducting water resources research is a strategic planning function: to make deliberative judgments about the allocation of funds and scope of research; to minimize duplication where appropriate (although sometimes more than one agency approach to the same problem can be productive); and to present to Congress and the public a coherent strategy for federal investment. Further, coordination can encourage more interdisciplinarity in the framing and conduct of research. In the absence of coordination, other more conventional activities can still add substantial value to current research management, such as more effective leveraging of research methods and capabilities. For example, agencies could increase the use and value of existing field research facilities by jointly sponsoring field experiments, technology development, and management of demonstration projects. More widespread use of interagency personnel exchanges would improve understanding of the missions and goals of other agencies.

National Agenda Setting and Strategic Planning

Chapter 3 introduced thinking about public investments in research as being analogous to a diversified financial portfolio, which is built on the premise that a diverse mix of holdings is the least risky way to maximize return on investments. In the context of water resources, a diversified research portfolio would capture the following desirable elements of a national research agenda: it would have multiple national objectives related to increasing water availability, to understanding water use, and to strengthening institutional and management practices; it would include short-, intermediate-, and long-term research goals supporting national objectives; the research would encompass agency-based, contract, collaborative, and investigator-driven research; it would address national and region-specific problems; and data collection would be in place to support all of the above.

In practice, the President and Congress implicitly define the balance among all these elements and subelements through the annual budget and appropriations processes. However, they are unable to do so explicitly because they lack (1) information about the size and shape of the whole portfolio, (2) measures of the individual research elements, (3) a consensus view of national priorities, and (4) guidance on what might constitute a productive balance of research elements. **Thus, the goal of coordinating water resources research is to enable the collection of information about the level and types of research and to advise OMB and Congress on a preferred shape of the entire portfolio, and in particular, a long-term research agenda to address national priorities in water resources.**

Note that a well-conceived vision of national priorities would not require revision each year; every three to five years would likely suffice. However, whether all the elements of the research portfolio are being adequately worked on would need to be assessed at least biannually to provide a basis for subsequent adjustments. This is not a trivial task. As described in Chapter 4, simply characterizing the dimensions of the research portfolio is very difficult to do under current budgeting practices. Furthermore, once the portfolio is characterized, appropriate performance measures need to be defined, capturing for example differences between short- and longer-term research. Indeed, performance metrics for research portfolios are now an active area of engagement among many federal agencies, stimulated in large measure by the Government Performance and Results Act.

To perform these functions, the coordinating body would bring together agency perspectives, an interdisciplinary perspective from the technical community, and a perspective that overarches the missions of the agencies. The latter point is critical. Unlike most areas of research, a significant portion of federally funded water resources research is conducted by scientists within the agencies, not by the external university-based or industry-based research community. As a consequence, the current portfolio of research, such as it is, conforms largely to the bounds set by agency missions, which may or may not comprehensively address national needs.

Data Sharing and Technology Transfer

The setting of data standards and the facilitation of data sharing are among the most critical value-added functions of coordination. One of the best examples of these functions is the Federal Geographic Data Committee (FGDC). The FGDC has brought together federal agencies, state and local governments, and the private sector to set numerous and sometimes complex standards for the collection, access, display, and storage of geospatial data including topographic, hydrographic, transportation, and cadastral (i.e., property boundaries) data layers (FGDC, 2004). Similarly, the Advisory Committee on Water Information and related activities such as the National Water Quality Monitoring Council promote compatible methods and standards for data to facilitate the sharing of data among agencies and as a consequence have improved data utilization.

Coordination also can facilitate technology transfer from research organizations to user communities. Over the last several decades, coordination mechanisms through the USGS Federal-State Cooperative Program and the National Water Quality Assessment program have been effective vehicles for conveying state-of-the-art surface water and groundwater modeling tools, water quality monitoring methods, and water management approaches.

A BRIEF REVIEW OF COORDINATION OF FEDERAL WATER RESOURCES RESEARCH

Over the last half century, Congress has occasionally opted for *temporary*, stand-alone bodies—most notably, the National Water Commission (see Chapter 2)—to consider national water issues and suggest both policy and research agendas. In retrospect, these mechanisms have had relatively little influence on the national research agenda. First, they rarely have had a constituency among the chief architects of agency budgeting. Second, there rarely is a ready-made implementing body to translate recommendations into tangible actions. Similarly, there are weaknesses associated with *permanent*, stand-alone coordination bodies. They are relatively easy to undercut, ignore, and disband. The demise in 1981 of the Water Resources Council illustrates this point and offers the further lesson that the operations of permanent coordinating bodies need to be carefully circumscribed to avoid political entanglements. Finally, Congress (or federal agencies) has often turned to the NRC, as is the case with this committee, to articulate a national water research agenda as viewed largely by the research community (e.g., NRC, 1981, 1991, 2001). There is evidence that some of these NRC efforts have had an impact on budgeting and program focus within agencies. For example, the report on the state of the hydrologic sciences (NRC, 1991) led to the creation of a new grants program administered by the National Science Foundation (NSF). However, because of the sporadic nature of such counsel, follow-up on recommendations and assessment of outcomes is difficult and, consequently, rarely done.

For the last 20 or so years, coordination of water resources research has occurred largely as an occasional exercise among the federal agencies and as an advisory activity to the Office of Science and Technology Policy (OSTP), with OMB staff involved but not in a leadership role. The basic concept is for agency representatives to get together to discuss emerging needs and share information about current programs. Even within this limited definition, coordination among agencies has occurred only sporadically over the last several decades, despite repeated calls for more coordination among the agencies (see Chapter 2).

After a hiatus of several years, OSTP has reconstituted an interagency group to examine water resources research priorities and needs. Since May 2003, the Subcommittee on Water Availability and Quality (SWAQ) of the National Science and Technology Council (NSTC) has been meeting on a regular basis. SWAQ has articulated an ambitious mission and agenda, included as Appendix E, built on the observation that the nation's research needs have fundamentally shifted with the advent (in the mid 1990s) of the "Third Era" of water resources, characterized most notably by a focus on in-stream water needs, surface water and groundwater interactions, and demands for improved biological and other monitoring needs. In general terms, a concept paper, written by the SWAQ's co-chairs and included in Appendix E, outlines potential areas of cooperation among agencies. According to its charter, the SWAQ may seek and receive advice from the President's Council of Advisors on Science and Technology (PCAST) and other outside groups. To date, that interaction has not occurred. It remains to be seen whether sufficient incentives are in place for SWAQ to realize its ambitious agenda, particularly pertaining to recommending budget priorities that could lead to reallocation of funding among agencies. As currently configured, SWAQ has few resources and staff to do much beyond analysis of gaps in specific research areas.

At the present time, coordination of a national water resources research agenda among the federal agencies and other interested parties, including the states and industry, is ad hoc and fragmented, to the extent that it exists at all. If agency missions were sufficiently inclusive and broadly viewed, interagency coordination would certainly suffice. However, the committee's view is that agency efforts to look beyond their own missions and define long-term national research priorities are necessary but not sufficient. Indeed, Congress and OMB closely scrutinize agency budgets specifically to avoid so-called "mission creep." The absence of a sustained, independent, broad, and long view of water research priorities means that both the administration and Congress are deprived of vital information to guide funding priorities.

OPTIONS FOR IMPROVED ANALYSIS, STRATEGIC PLANNING, AND COORDINATION

Based on these past and ongoing experiences, the committee has concluded that an effective and sustainable coordinating body needs to draw from a constituency or community of experts that goes beyond the agencies themselves, but that is integrated to the extent possible into existing processes. If the coordinating body is made up only of agency representatives, the overarching national perspective will likely devolve to the sum of agency wish lists. However, independence from agency agendas needs to be balanced by close interaction with agency leaders who have unique and valuable perspectives on national needs as seen through the lens of their missions. Further, agencies need to feel that they have a positive stake in the outcome from coordination and not simply and reflexively assume a defensive posture. Thus, the coordinating body will need to strike a balance between independence and integration into existing institutions.

The coordinating body should have a clear mandate from Congress, which plays a critical role in the agenda-setting and funding process. Congress conferring legitimacy on a coordination process elevates its importance within the executive branch and its relevance to outside constituencies. The coordinating body also needs a reliable means to tap into stakeholder groups and other constituencies to learn of their needs, and to communicate potential new directions for which feedback is desired.

Past experiences also suggest that an effective coordination mechanism should be synchronized with the schedule of the federal budgeting and appropriations processes to maximize impact. An effective coordination mechanism should be cyclical and sustainable in order to provide the flexibility that will be needed to address future unknown problems. During "on years," the coordination body could focus on adjusting the national research portfolio. During "off years," the focus would be on assessing the effectiveness of implementation, thus allowing a determination of the value added by the coordination effort. Sustainability derives from a demonstrated ability to add value to the agenda-setting and budgeting process.

Several options were considered to provide coordination among the multiple research and user communities and advise the Congress and OMB on the key long-term priorities of a national water resources research agenda. Each of the options discussed below would increase the likelihood that at least some of the basic functions of data collection, information sharing, and national priority setting might be implemented.

Option 1: Existing NSTC Subcommittee

Option 1 is a slight variation on the status quo as of this writing. The NSTC was formed in 1993 as a successor to the Federal Coordinating Committee for Science, Engineering, and Technology. Members of the council include almost every cabinet secretary and major agency head. Beneath the NSTC are several standing committees whose members include senior leadership from the agencies. The relevant NSTC committee for water research is the Committee on Environment and Natural Resources (CENR); a subcommittee within CENR is devoted to water issues.

For most of the last decade, the water resources research subcommittee has been dormant. However, as described earlier, the SWAQ was revived in 2003 and appears to be functioning effectively as a forum for agency representatives to share information about their respective programs. The SWAQ's activities have not yet extended beyond information exchange among agencies, although its charter explicitly calls for the SWAQ to provide advice on national agenda setting.

As of this writing, the SWAQ plans to release two reports in 2004: an overview of water availability and use that identifies knowledge gaps, and a report on the linkage between land use and water quality.

This coordination option has its attractions. Arrangements are already in place, and agency roles and responsibilities are well defined. In the past, this mechanism has been used to direct and analyze a "data call" from OMB to the agencies when specific budget and program information has been sought to support an administration initiative. In principle, this function could be expanded to collect consistent and comparable information from the agencies every two years or so about the nature and extent of their research activities (for example, in a manner similar to the survey in Box 4-1); the effort could be timed to coincide with the annual budgeting and appropriations processes.

This option also could include a competitive grants program located within the NSF. A competitive grants program would serve two important functions: to increase the proportion of long-term research and to address topical gaps in the current water resources research portfolio. This program would require new (but modest) funding. To be effective in meeting its purposes and to address the gaps noted in Chapter 4, **funding would be needed on the order of $20 million per year for research related to improving the efficiency and effectiveness of water institutions, and $50 million per year for research related to challenges and changes in water use.** Along with everyone else, scientists within the federal agencies would be allowed to compete for those funds and indeed would have an incentive to demonstrate their capacity to conduct interdisciplinary and systems-based research. This would provide existing agencies with a positive incentive to participate in a program of this sort, although it represents a departure from current practice. The competitive grants program would give Congress and OMB latitude to pursue new lines of research without necessarily disrupting mission-driven programs requiring sustained, long-term funding.

Option 1 also has significant shortcomings. After many iterations during previous administrations, this approach has yet to demonstrate that it can be an effective forum for looking beyond agency missions to fundamental research needs. In the absence of new funding, the tendency of program managers is to protect their agencies' interests. Incentives and rewards for agency-to-agency coordination and higher-level agenda setting are usually too meager to merit attention by any means. To make this mechanism more effective than the current NSTC apparatus, the OMB budget coordination function would need to be strengthened and made explicit in the charter of the SWAQ. Otherwise, agencies would have little incentive to participate in any meaningful way or abide by recommendations that might have an adverse effect on their own budgets. Further, without new funding, the SWAQ would not have the resources or the staff needed to actually carry out a budget data call and subsequent analysis.

Another weakness of this approach is that the SWAQ lacks connections—formal or informal—to states, stakeholders, and other users. Few members of

Congress know of its existence. Its primary audience is OMB and the agencies. As such, the SWAQ is invisible to the public at large as well as the research community outside of the federal agency leadership; as yet, it has conducted no outreach activities.

Option 2: A "Third Party" Water Research Board Model

A second option involves Congress authorizing a neutral third party called the Water Research Board to carry out the following functions:

- do a regular survey of water resources research using input from federal agency representatives
- advise OMB and Congress on the content and balance of a long-term national water resources research agenda every three to five years
- advise OMB and Congress on the adequacy of mission-driven research budgets of the federal agencies
- advise OMB and Congress on key priorities for fundamental research that could form the core of a competitive grants program administered by NSF or a third party (identical to that described above under Option 1)
- engage in vertical coordination with states, industry, and other stakeholders, which would ultimately help refine the agenda-setting process

In contrast to Option 1, a Water Research Board would place the outside research and user communities on equal footing with agency representatives. An advantage of this mechanism is that a national research agenda would reach beyond agency missions to include the views of broad-based research and user communities.

The Water Research Board could exist, for example, as a standing committee or commission consisting of prominent individuals, with term limits, from one or more national professional scientific societies or trade associations concerned with water resources research (e.g., the American Society of Civil Engineers, the American Geophysical Union, Sigma Xi, the Water Environmental Foundation). It could be organized and operated in a manner similar to which the NRC selects and manages its standing boards and committees. The Water Research Board would ideally report to OMB, OSTP, and the Congress on a cycle compatible with the budget and appropriations processes. In return, OMB would gain a credible source of advice on a regular basis that it could use to improve the efficiency with which the federal agencies fund and conduct water resources research. Congress would similarly be assured that the advice being given to OMB integrated the interests of a community that extended well beyond the agencies. Agencies would have the opportunity to make a case for their own research agendas and build a constituency beyond their traditional interests. This option, of the three proposed, is the most likely to provide a credible and unbiased view of research needs and priorities to Congress and the administration. Further, the independence from the agencies afforded by this option makes it possible to focus the competitive grants program on longer-term research needs, particularly those needs falling outside the areas of interest of the agencies such as water use and water institutions.

A disadvantage of Option 2 is that it may engender resentment from the agencies by requiring their regular participation in the survey exercise and if it draws resources away from the agencies' mission-driven research. Indeed, as with the SWAQ under Option 1, a Water Research Board would require modest funding both for the competitive grants program and for its own operation. Ideally, this would be provided by Congress, but it may be drawn from agency base budgets. It is also possible that such a mechanism would create pressure for new funding that OMB and Congress would prefer to avoid. Further, OMB may be reluctant to establish a formal advisory body, preferring instead to work within the NSTC framework and gather stakeholder views informally through normal, ad hoc channels. Finally, this option suffers from the fact that OMB would be under no obligation to take the advice given by the Water Research Board, thus limiting its potential effectiveness.

Option 3: OMB-led Model

A hybrid model of a Water Research Board described in Option 2 would be led by OMB and formally tied to the budget process. OMB is the only federal agency in a position to implement budget-based coordination and "crosscutting" programmatic functions. In this model, OMB would chair a committee of senior-level agency officials; a working group of staff to these senior officials could provide the background material and suggested agenda items for the senior-level group. This group could also establish and receive the views of a federal advisory committee made up of leading scientists, state and local officials, representatives from the business community, environmental groups, labor, and any other relevant stakeholders. However, at its root, this option is an OMB-led coordination mechanism that would place agency budget exercises at the center of its activities and most directly address congressional concerns about the coherence of the federal investment.

It is legitimate to ask why OMB should perform this function for water resources research and not for many other areas of research that cut across agency lines. While there clearly are other multiagency research areas, water resources research stands out as particularly in need of repair because of the sheer volume of agency players, the critical importance of water in the economy, and the history of fragmentation and compartmentalization of research activities by agency mission instead of by broader national and regional needs. Another reason for singling out water resources research as an area worthy of OMB's special attention is their inability to accurately estimate the magnitude and character of federal investment in this area or find another entity to do so on a sustained and unbiased basis. There is precedence for highly successful OMB-led coordination in the realm of the geosciences. OMB has played a leading role in implementing various integrated budget activities, like Bulletin 17, governing the estimation of flood frequency for use in evaluating proposed flood control projects and federal flood insurance (IACWD, 1982), and Circular A-16, governing the collection of geospatial data (OMB, 2002).

OMB works with the OSTP to review the research components of all of the agencies as a routine part of the budget process. In practice, active coordination of research and development budgets across agencies rarely occurs outside of a formal "crosscutting" exercise like USGCRP. Further, OMB staff participate in nearly all of the subcommittees and committees of the NSTC and PCAST. Both NSTC and PCAST are managed by OSTP.

In this option, the committee of senior-level officials—under OMB's direction—would perform the same functions as those listed in Option 2 for the Water Research Board. The difference is that in this case, OMB would run the process, so that it would feed more directly into their activities. Like Option 2, this option provides a structured process to solicit a wide range of views and then synthesize a national research agenda. It is tied to budget and policy processes, and it would be a credible convener of scientific, federal, state, and private sector interests. An OMB-led committee would require a modest budget and structure for agenda setting and for administering the competitive grants program, and as with both other options, agencies might end up footing the bill unless new money is appropriated. OMB and Congress would be assured of a credible infusion of advice on national priorities, with an integration of views from a much broader range of stakeholders than would be the case through the NSTC option described in Option 1.

However, OMB clearly would be challenged to manage the increase in staff and funding required to meet the objectives of a Water Research Board as described in Option 2. OMB's interest in rising to the challenge would depend on perceived gains, from both a budget and program perspective. Those gains would not necessarily be obvious from the perspective of an agency budget examiner. Hence, for Option 3 to be viable, support for the process would need to be strong and clearly communicated by senior management to budget staff.

A strength of Option 3—its close connection to the budgeting process—is also a potential weakness because of conflicts of interest. Depending on its agenda at any given time, OMB may not have an interest or incentive to give the committee of senior-level agency officials free reign to advise on a research agenda in an objective manner in the same way that an independent, external committee could (Option 2), but rather might insist that the committee of senior-level agency officials hew to an administration position. In other words, the quest for an unbiased accounting of national water research needs might not necessarily align with other objectives such as balancing the budget or supporting other lines of research. A congressional mandate and ongoing oversight could partially mitigate this potential shortcoming.

ISSUES OF VERTICAL COORDINATION

Regardless of the chosen mechanism for coordination of water resources research, it will be important to ensure two-way communication between the generators and sponsors of research (primarily the federal agencies) and the users of such research (e.g., state and local governments). In this regard, it is worth mentioning the potential role for the Water Resources Research Institutes in helping achieve this vertical coordination. The Water Resources Research Institute system, described in Chapter 2, provides an existing, well-organized mechanism for articulating state-based research needs and for bringing together water managers, stakeholders across a wide cross section of the public, and academic researchers and academic institutions throughout each state. Each institute convenes advisory groups of managers, academics, and stakeholders to formulate and periodically review the research agenda of that state, and it uses a peer-review system for competitively awarded research funds. The institutes also are required to maintain an information transfer program, which links research with the user public in a two-way flow of information. Because the institute system has been in existence for nearly 40 years, these linkages are well established. The institute system can provide an effective means of communication between, for example, a national-level Water Research Board and the state and regional water resources research needs. A review of the Water Resources Research Institutes completed five years ago (Vaux, 2003) corroborates the view that the institutes are capable of shouldering the task of interlocutors between state and local research needs and federal agencies and funders.

CONCLUSIONS AND RECOMMENDATIONS

Coordination of water resources research is more than gathering representatives of federal agencies together in a room several times a year to exchange notes. As conceived here, coordination includes national research agenda setting, leadership in tackling large and complex multiagency research efforts targeted at emerging problems, sustained attention to the composition of the research portfolio and identification of gaps, and a competitive grants program that addresses national research needs unaddressed by agency missions.

Coordination of the water resources research enterprise is needed to make deliberative judgments about the allocation of funds and scope of research, to minimize duplication where appropriate, to present Congress and the public with a coherent strategy for federal investment, and to facilitate the large-scale multiagency research efforts that will likely be needed to deal with future water problems. This report has documented the need for coordination to reach beyond agency-defined research priorities to a broader community of interest and expertise in national water research. A data collection effort that would enable Congress and OMB to understand the full dimensions of the federal water resources research portfolio is also a necessity. Finally, the case has been made for a competitive grants program focused on heretofore ignored national water resources research priorities.

Any one of the three options presented in this chapter could be made to work in whole or part to meet these needs; each has strengths and weaknesses that would need to be weighed against the benefits and costs that could accrue from moving beyond the status quo. Option 1 has the distinct advantage of currently being in existence and having the full support of OMB and OSTP. However, the NSTC process lacks a historical record of success, visibility outside of the executive branch, and support beyond the participating agencies. By definition, its role is limited by its composition and its narrowly scoped mission. From a budget perspective, the NSTC process would be an unlikely source of advice to Congress and OMB on reallocating existing water resources expenditures.

As described in Option 2, a Water Research Board associated with an objective, third-party organization would represent a marked improvement in representation and consensus-building beyond the Option 1 mechanism. Further, the independence from the agencies afforded by this option makes it possible to focus the competitive grants program on those research needs falling outside the areas of interest of the agencies. But it is possible that what would be gained in breadth, independence, and objectivity would be lost in detaching the coordination mechanism from existing government arrangements. Unless explicitly directed otherwise by Congress, OMB might attach no particular imperative to the Water Research Board's recommendations.

Option 3 has the advantage of being an OMB-led venture from the start, and it would be naturally tied to routine executive branch schedules for budget development and review. As with Option 2, OMB would have the advantage of an expert and representative body that would enable it to tackle an area of the budget that has hitherto escaped an effective national focus. However, an OMB-led effort would lack the independence that would be provided in Option 2 without overcoming the financial hurdles, and it might suffer from internal conflicts of interest.

In the end, decision makers will choose the coordination mechanism that meets perceived needs at an acceptable cost in terms of level of effort and funding. It is possible that none of the options is viable in its entirety. However, it may be possible to partially implement an option, which in itself would be an improvement over the status quo. For example, the initiation of a competitive grants program targeted at high-priority but underfunded national priorities in water resources research could occur under any one of the options and in lieu of the other activities listed above.

The coordination problem—broadly conceived—is eminently solvable, but it is unlikely to be solved without a concerted effort by leaders in Congress and the administration. However, with the strategic application of national-level leadership, modest resources, and a sharp focus on national water resources research needs for the 21st century, the opportunity for substantially improving on the status quo is within reach.

REFERENCES

Federal Geographic Data Committee (FGDC). 2004. The Federal Geographic Data Committee: Historical Reflections—Future Directions. An FGDC paper on the history of the FGDC and NSDI. January 2004. http://www.fgdc. gov.

IACWD (Interagency Advisory Committee on Water Data). 1982. Guidelines for Determining Flood Flow Frequency. Bulletin 17–B. Reston, VA: USGS Office of Water Data Coordination.

National Research Council (NRC). 1981. Federal Water Resources Research: A Review of the Proposed Five–Year Program Plan. Washington, DC: National Academy Press.

National Research Council (NRC). 1991. Opportunities in the Hydrologic Sciences. Washington, DC: National Academy Press.

National Research Council (NRC). 2001. Envisioning the Agenda for Water Resources Research in the Twenty–First Century. Washington, DC: National Academy Press.

National Science and Technology Council (NSTC) Committee on Environment and Natural Resources. 2003. An Assessment of Coastal Hypoxia and Eutrophication in U.S. Waters. Washington, DC: Executive Office of The President of the United States.

Office of Management and Budget (OMB). 2002. Circular No. A–16: Coordination of Surveying, Mapping, and Related Spatial Data Activities. http://www.whitehouse.gov/omb/circulars/-a016/a016_rev.html.

Vaux, H. J., Jr. 2003. Results of the 1999 Evaluation of the State Water Resources Research Institutes mandated by the Water Resources Research Act of 1984 (P.L. 98–242).

APPENDIXES

Appendix A

Modified FCCSET Water Resources Research Categories

All of these categories apply to the in-house work of the agencies as well as what the agencies fund externally.

I. NATURE OF WATER

Category I deals with fundamental research on the water substance.

A. Properties of water—Study of the physical and chemical properties of pure water and its thermodynamic behavior in its various states.

B. Aqueous solutions and suspensions—Study of the effects on the properties of water of various solutes, surface interactions, colloidal suspensions, etc.

II. WATER CYCLE

Category II covers research of a basic nature on the natural processes involving water. It is an essential supporting effort to applied problems in later categories. Routine sampling and data collection are excluded.

A. General—Studies involving two or more phases of the water cycle such as hydrologic models, rainfall-runoff relations, surface water and ground-water relationships, watershed studies, etc. *Note: This subcategory generally includes research activities at the subcontinental scale or smaller. Water cycle studies on a global scale should be included under II-M.*

B. Precipitation—Investigation of spatial and temporal variations of precipitation, physiographic effects, time trends, extremes, probable maximum precipitation, structure of storms, etc. Research on temporal variation in precipitation at decadal and longer scales and the influence of ocean processes in driving this variability would also be in this subcategory. This subcategory also includes studies of how changes in land use/land cover affect precipitation.

C. Snow, ice, and frost—Studies of the occurrence and thermodynamics of water in the solid state in nature, on spatial variations of snow and frost, on formation of ice and frost, on breakup of river and lake ice, on glaciers and permafrost, on the theory and use of satellites for monitoring snow cover, etc.

D. Evaporation and transpiration—Investigation of evaporation from lakes, soil, and snow and of transpiration in plants; methods for estimating actual evapotranspiration; and energy balances. Research on new space approaches for monitoring atmospheric fluxes of water vapor with particular emphasis on evaporation and transpiration would be included.

E. Streamflow and runoff—Study of the mechanics of flow in streams and overland flow; flood routing; bank storage; space and time variations (includes high- and low-flow frequency); droughts; and floods. *Note: Studies dealing primarily with the natural chemical or biological aspects of streams should be incorporated in II-K and XI-A, respectively. Studies concerned with contamination of streams should be included in V-B.*

F. Groundwater—Study of the mechanics of groundwater movement; multiphase systems; sources of natural recharge; mechanics of flow to wells and drains; subsidence and other properties of aquifers; etc. *Note: Studies dealing primarily with the natural chemical or biological aspects of groundwater should be incorporated in II-K and XI-A, respectively. Studies concerned with the fate and transport of pollutants in groundwater should be included in V-B.*

G. Water in soils—Studies of infiltration, movement, and storage of water in the zone of aeration, including soil. This section should include new technologies that are evolving for monitoring soil moisture, for example using space-based platforms.

H. Lakes—Study of the hydrologic, hydrochemical, and thermal regimes of lakes; water level fluctuations; currents and waves. This section should include information on the natural physical and chemical processes that affect lake-water chemistry. *Note: Studies dealing primarily with biological aspects of lakes should be in XI-A, while those involved with the fate and transport of contaminants in lake systems should be considered in V-B.*

I. Water and plants—Understanding the role of plants (including crops) in the hydrologic cycle; water requirements of plants; interception. *Note: Studies of water conservation by using more water-efficient plants in agriculture should be in III-F, while studies on controlling the growth on phreatophytes should be in III-B.*

J. Erosion and sedimentation—Studies of stream geomorphology; the erosion process; prediction of sediment yield; sedimentation in lakes and reservoirs; stream erosion (aggradation and degradation); sediment transport; sheet and rill erosion from farm, forest, range, and other land uses; relationship between urbanization and sedimentation; etc.

K. Chemical processes—Understanding chemical interactions between water and its natural environment; chemistry of precipitation. Include both surface water and groundwater studies.

L. Estuaries—Studies dealing with physical/chemical processes and special problems of the estuarine environment, including the effect of tides on flow and stage, deposition of sediments, and seawater intrusion in estuaries. *Note: Studies dealing primarily with biological aspects of estuaries should be incorporated in XI-A. Studies involved with the fate and transport of contaminants in estuaries should be considered in V-B.*

M. Global water cycle problems—Studies by U.S. researchers on global water cycle problems (including those brought about by climate change) that influence water resources in the United States. Examples might include estimates of river discharges to oceans on a worldwide basis, and effect of man-made impoundments or land-use changes on the global water cycle. *Note: Studies that focus primarily on the effects of climate change on aquatic ecosystems should be reported in XI-C.*

III. WATER SUPPLY AUGMENTATION AND CONSERVATION

As water use increases we must pay increasing attention to methods for augmenting and conserving available supplies. Category III is largely applied research devoted to this problem area. *Note: There is overlap with other categories. For example, pollution control and treating impaired waters for reuse (V) may serve to augment available supplies. Better planning and management of water resources (VI) and improved engineering works (VIII) may also have the effect of increasing the utility of water resources. Only if the primary goal of the project is to*

***augment supplies** should it be classified in III. If the primary objective is to control flow, it should go under IV.*

A. Saline water conversion—Research and development related to methods of desalting seawater and brackish water.

B. Water yield improvement—This subcategory embraces research on a variety of techniques to recover some part of the water returned to the atmosphere by evaporation or transpiration or other losses. This includes studies on increasing streamflow or improving its distribution through land management; determining hydrologic response to artificially induced rainfall; water harvesting from impervious areas; phreatophyte control; reservoir evaporation suppression.

C. Use of water of impaired quality—Research on using water of impaired quality. This includes, for example, understanding the agricultural use of water of high salinity (and issues of crop tolerance to salinity), the use of poor-quality water in industry, and the use of "gray" water by municipalities. *Note: Treatment of impaired waters for subsequent reuse as a higher-quality source is considered in V-D.*

D. Conservation in domestic and municipal use—Developing methods for reducing domestic water needs without impairment of service; demand management. (This does not include temporary water restrictions imposed during droughts.)

E. Conservation in industrial use—Developing methods for reduction in both consumption and diversion requirements for industry.

F. Conservation in agricultural use—Studies on more efficient irrigation practices, including chemical control of evaporation and transpiration; development of lower-water-use plants and crops; methods of applying irrigation and timing issues; methods for control of deep seepage; etc.

IV. WATER QUANTITY MANAGEMENT AND CONTROL

Category IV includes research directed to the management of water, exclusive of conservation, and the effects of related activities on water quantity. Routine sampling and data collection are excluded. *Note: The choice between II and IV is dependent on whether the research is directed toward understanding basic processes (II) or evaluating man's control efforts (IV). Also, it is recognized that water quantity management and control (IV) can result in changes in water quality.*

A. Control of water on the surface—Study of the effects of land management on runoff (e.g., reshaping of the land surface by terraces, constructed wetlands, and other structures and by extending the role of vegetation in influencing runoff); land drainage; seepage control; effect of control programs and devices on the stage and time distribution of streams, lakes, and estuaries; control of noxious weeds and objectionable plant growth in surface channels.

B. Groundwater management—Study of artificial recharge of groundwater aquifers, conjunctive operation, and their relation to irrigation. *Note: Research in this subcategory is supported by, but should be differentiated from, more general research on groundwater (II-F), chemical processes (II-K), and reclaimed wastewater (V-D) (which may be a useful source for groundwater recharge).*

C. Effects on water of man's non-water activities—Understanding the impact of urbanization, transportation systems, logging, grazing, mining, traditional agriculture, etc., on water yields and flow rates (that is, the quantity and time distribution of water). *Note: This subcategory is different from IV-A because it focuses on basic understanding of effects, not on management actions.*

D. Watershed activities—This subcategory refers narrowly to methods of controlling erosion to reduce sediment load of streams and to conserve soil. *Note: The broader connotation of watershed protection that includes research on structural management measures (BMPs) that may protect water resources from detrimental changes in water quantity and quality is found in IV-A and V-G, respectively.*

V. WATER QUALITY MANAGEMENT AND PROTECTION

Increasing quantities of municipal, industrial, agricultural, and other wastes containing physical, chemical, and biological pollutants are entering surface waters and groundwater. Category V deals with methods of identifying, describing, and controlling this pollution. Included also are studies on the fate of contaminants in the environment and the effects of contamination on various uses of water resources. Routine sampling and data collection are excluded.

A. Identification of pollutants—Development of techniques for identification, detection, and quantification of physical, chemical, and biologic pollutants in water. This should encompass both traditional pollutants as well as emerging contaminants like toxins, microbial pathogens, pharmaceutical compounds, pesticide metabolites, and other compounds in trace amounts that may have reproductive or endocrine effects and affect human and aquatic health in diverse ways.

B. Sources and fate of pollution—Research to determine the sources of pollutants in water, the nature of the pollution from various sources, the fate and transport of pollutants from source to stream or groundwater, and the transformation of pollutants due to physical, chemical, or biological action.

C. Effects of pollution on water resources—Studies of the effect of pollutants (singly or in combination) on different uses of water: municipal, industrial, agricultural, recreational, and propagation of aquatic life and wildlife (as sometimes assessed via ecological risk assessment). This also includes studies on the cause of eutrophication in fresh and marine waters. *Note: Research on how waterborne pollutants affect human health should be reported in V-H.*

D. Waste treatment processes—Research on physical, chemical, and biological treatment processes to remove or modify impurities in wastewater, including new types of waste; research on treatment methods for more complete removal of pollutants that will facilitate water reuse for domestic, agricultural, or industrial purposes; research on treatment processes for remediating contaminated groundwater.

E. Ultimate disposal of wastes—Studies on the disposal of residual material resulting from treatment of contaminated waters. Such wastes include the material removed from municipal, industrial, and agricultural waters during treatment, the waste brines from desalination or oil fields, radioactive waste concentrates, wash waters from filters at drinking water treatment plants, etc. This includes disposal to both land areas and receiving waterbodies including groundwater.

F. Water treatment and distribution—Development of more efficient and economical methods of water treatment for municipal, industrial, agricultural, or recreational uses. This includes alteration of water quality to prevent deterioration during storage and distribution.

G. Water quality control—Research on methods to control surface water and groundwater quality by all methods (except waste treatment, which is in V-D) such as production modification or substitution, process changes, improved agricultural practices for preventing pollution from pesticides and other agricultural chemicals; other structural management measures (BMPs) to protect water resources from water quality changes brought about by various land-use practices (nonpoint source pollution control).

This also includes management of waters to improve water quality such as flow augmentation and supplemental aeration. Thus, it complements IV-A, IV-D, and VI-F.

H. Effects of waterborne pollution on human health—Chemical and microbial risk assessment methods development and underlying research such as statistical, exposure, and dose-response studies. This includes both laboratory and field research that contributes knowledge to human health risk assessments for both general and susceptible subpopulations (e.g., children, the elderly, immune-compromised groups).

VI. WATER RESOURCES PLANNING AND OTHER INSTITUTIONAL ISSUES

The problems of achieving an optimal plan of water development are becoming increasingly complex. Category VI covers research devoted to determining the best way to plan, the appropriate criteria for planning, and economic, legal, and institutional aspects of water resources management, as well as other contributions from the social sciences. It does not include economic, legal, and social analyses that represent an integral phase of research activities conducted under the other major categories.

A. Techniques of planning—Application of systems analysis to project planning; treatment of uncertainty; probability studies; decision support systems.

B. Evaluation process—Development of methods, concepts, and criteria for evaluating project costs and benefits; estimating project life and other economic, social, and technological parameters into the future; research on discount rates and planning horizons.

C. Economic issues—Studies of the role of prices and markets in managing water resources (including price and income elasticities, water markets and quasi-markets, the value of water in alternative uses, the costs of providing water via different technologies, and techniques of cost allocation); cost sharing; and repayment policy. This subcategory includes research on economic methodologies that may increase the effectiveness and breadth of water-related economic institutions.

D. Water demand—Research on the water quantity and quality requirements of various water-use sectors, both diversion and consumption, excepting environmental uses (in XI-A).

E. Water law and institutions—Studies of state and federal water laws, including groundwater laws, with an emphasis on changes and additions that will encourage greater efficiency in use and other issues. This subcategory also includes investigation of institutional structures and constraints that influence decisions on water at all levels of government.

F. Nonstructural alternatives—Exploration of methods to achieve water development aims by nonstructural methods. This includes research on nonstructural practices that might be part of water resources and watershed management, including land acquisition, conservation easements, zoning and development ordinances, septic system siting requirements and inspection, economic incentives, incentives to reduce impervious cover, and public education and outreach.

G. Risk perception and communication—Research focused on the ways in which people perceive water-related risks and how that impacts public decision making, as well as alternative ways of communicating about such risk.

H. Other social sciences—Studies in anthropology, geography, political science, psychology and sociology related to water. Such studies include those focused on different value structures and cultural norms, experience with processes for obtaining stakeholder input, and cultural, ethical, and religious facets associated with water and its use.

I. Water resources policy—Research focused on the determinants of water resources policy and on methods for generating policies based on sound scientific concepts. This subcategory also includes *ex post* studies of how past projects and policies have performed.

VII. RESOURCES DATA

Category VII includes research oriented to data needs and the most efficient methods of meeting these needs. This category contains only that research primarily devoted to improving strategies for establishing field data collection programs, data acquisition methods, and data evaluation, processing, and dissemination. (This includes waterborne disease surveillance systems.) It does not include research on new monitoring techniques for individual parameters (such research would be categorized in II, III, IV, V, or VI). Finally, no funds reported in this category are for data collection as such.

A. Network design—Studies of data requirements and of the most effective methods of collecting the data.

B. Data acquisition—Research on new and improved instruments and techniques for collection of water resources and related data, including automation, telemetering equipment, and remote sensing.

C. Evaluation, processing, and publication—Studies on effective methods of processing and analyzing data, on compiling such data and information into databases, and on the most effective form and nature of published data, including web sites and maps.

VIII. ENGINEERING WORKS

To implement water development plans requires engineering works. Category VIII describes research that has as its prime objective the development of improved technology for designing, constructing, and operating these works. Works relevant to a single specific goal, such as water treatment or desalination, are included elsewhere if an appropriate category exists.

A. Structures—Research on the design and construction criteria and techniques for all structures associated with the development of water resources or the control of surface water or groundwater. Included are dams, locks, bridges, conduits, lined tunnels, floodwalls and levees, water supply intakes, wells, pipelines, and reservoirs.

B. Hydraulics—Studies of the static and dynamic behavior of water as it influences design theory for spillways, penstocks, conduits, tunnels, canals, rip-rap, breakwaters, floodwalls, wells and well systems, and other similar structures.

C. Hydraulic machinery—Design and performance of hydraulic machinery and equipment including gates, valves, pumps, turbines, and similar facilities. Includes associated control facilities, generators, transmission systems, and power system operation to the extent each is unique to problems of water utilization.

D. Soil mechanics—Research on the design theory, criteria, techniques, and engineering properties of soils as related to the design, construction, and performance of cut slopes, earth foundations for water resources engineering works, embankments, and rockfill structures.

E. Rock mechanics and geology—Study of the behavior of rock masses and rock foundations, engineering characteristics, and structural properties of rock materials, and design techniques applicable to foundations for large structures—for application specifically to water resources engineering works.

F. Concrete—Research on cementing materials, aggregates, and other concrete components; engineering characteristics of concrete construction methods and techniques that pertain to water resources engineering works.

G. Materials—Study of miscellaneous materials other than soil, rock, concrete, and concrete components, and study of detection, measuring, and materials testing techniques and equipment. Areas included are bituminous materials, chemical materials, metals, paints and corrosion, and plastics or synthetic materials associated with water or structures for water control.

H. Rapid excavation—Research on the mechanical, chemical, and nuclear explosive techniques and equipment for rapidly excavating and moving large volumes of earth or rock—for application specifically to water resources engineering works.

I. Fisheries engineering—Research and development of techniques and design of facilities to attract and pass fish past dams and other water control structures. Development of methods for improving the design, maintenance, and functioning of fish spawning areas.

J. Infrastructure repair and rehabilitation—Research on how to extend the life and maintain the safety of existing water structures (e.g., dams, pipelines, aqueducts, etc.).

K. Restoration engineering—Research on the restoration of natural flows and mitigation of ecological impacts via reversal of engineering works, including dam removal, modification of existing structures, etc.

L. Facility protection—Research on structural and nonstructural means of protecting water-related facilities from terrorism or other threats.

IX. MANPOWER, GRANTS, AND FACILITIES

Trained personnel are an essential ingredient of research on water resources and the planning and design of water development projects. Category IX describes plans for support of education and training. It also includes grant and contract programs for which allocation to other categories is impossible.

A. Education—extramural—Support of education in water resources and water-related fields at universities. This includes training and facilities grants in the field of water resources, but it does not include grants made under any of the other research categories listed above.

B. Education—in-house—Government employee training programs.

C. Capital expenditures for research facilities—This subcategory includes estimates of the capital cost of research laboratories, field stations, etc., needed in the current fiscal year.

D. Grants, contracts and research act allotments—Allotments to University Water Resources Research Institutes under P.L. 88-379; OWRT, HEW, NSF, CSRS, and other grants that cannot be distributed to other categories.

X. SCIENTIFIC AND TECHNICAL INFORMATION

Development of adequate manual or mechanized procedures for acquisition, storage, retrieval, and dissemination of scientific and technical information is a vital and integral part of a successful research program. This category includes all separately identifiable activities involved in the handling of recorded knowledge resulting from basic or applied research in the water-related aspects of the physical, life, and social sciences. Such activities include the planning and performance of functions related to:

A. Acquisition and processes—Identification, acquisition, storage, or exchange of documents in full size or reduced form, and the organization or arrangement of these documents for retrieval.

B. Reference or retrieval—Selected search and retrieval of an organized document collection in response to specific user request.

C. Secondary publication and distribution—Selective review, indexing, subject classification, coding, abstracting, announcing, listing, or distribution of documents or their bibliographic surrogates to provide services such as current awareness or selective dissemination of information systems, abstract bulletins, and topical bibliographies.

D. Specialized information center services—Activities described under subcategories A, B, and C where performed by a separate functional element whose mission includes additional subject area technical competence to critically review, digest, analyze, evaluate, or summarize scientific and technical information in specially defined areas, or to provide advisory or other services.

E. Translations—Conversion, to or from English, of scientific or technical documents in whole or part, where performed as a separate or specific activity.

F. Preparation of reviews—Preparation of state-of-the-art critical reviews and compilation in specified technical subject areas, where performed as a separate activity exclusive of the output of subcategory D.

XI. AQUATIC ECOSYSTEM MANAGEMENT AND PROTECTION

Increasing water demands for human use, the effects of changing land-use practices on receiving waters, an expanding list of contaminants, and climate change and alteration of global biogeochemical cycles challenge the effective management and protection of aquatic ecosystems. Biological diversity and ecosystem processes of lakes, wetlands, and rivers are increasingly at risk, with the potential degradation of ecosystem goods and services and loss of species. Most prior research has focused on a narrow view of water quality for human use and direct harm to sensitive species. Research under Category XI is directed to understanding aquatic ecosystem management and protection in coupled, complex systems, including studies of long duration and large spatial scale.

A. Ecosystem and habitat conservation—Research to gain improved understanding of the coupling of hydrologic and ecological processes, such as the ecological outcomes expected from particular flow regimes, hydroperiods, and geochemical conditions. It also includes research into all aspects of aquatic ecosystem structure and function, including water requirements of aquatic and related terrestrial ecosystems, ecosystem response to degradation, procedures for restoration, and response of the biota to management alternatives. *Note: Research on restoration strategies may overlap with IV-D, V-G, and VIII-K. The decision as to where to categorize activities should depend on the primary goal (i.e., water quantity control vs. ecosystem restoration).*

B. Aquatic ecosystem assessment—Research into methods and models to assess the status of aquatic ecosystems, and development of indicators and indices of ecological integrity to identify locations where restoration is appropriate, to provide a means of monitoring long-term trends in ecosystem status, and to quantify trends in improvement or deterioration over time in response to human actions. This subcategory includes efforts to develop metrics of the monetary value of ecosystem services.

C. Effects of climate change—Research to determine the complex direct and indirect pathways by which climate change will impact freshwater ecosystems and their biological productivity, including changes in water quantity and quality, biogeochemical cycles, and food webs.

D. Biogeochemical cycles—Research to understand and predict the cycles of C, N, P, S and other elements at the global scale; to understand the sources, fluxes, transformations, and fate of these elements; and to understand how humans have affected the global cycling of these elements and the resulting impacts on climate, biological production, and ecosystem processes. *Note: Applied research into the local fate and transport of pollutants including nutrients should be reported under V-C.*

Appendix B

Survey Data from Federal Agencies and Nonfederal Organizations

TABLE B-1 Funding Levels for Water Resources Research for all the Federal Agencies, and Levels for EPA, NSF, and DOI for FY1999–2001 (data in the thousands of dollars; not adjusted for inflation)

	All Federal Agencies			EPA		
Research Category	*1999*	*2000*	*2001*	*1999*	*2000*	*2001*
I. Nature of Water						
A. Properties of water	680	1087	1251			
B. Aqueous solutions	6919	10066	11008			
Subtotal	*7599*	*11153*	*12259*			
II. Water Cycle						
A. General	10635	14209	10290			
B. Precipitation	14369	14693	14831			
C. Snow, ice, and frost	5090	4964	5122			
D. Evaporation and transpiration	8590	9390	9498			
E. Streamflow and runoff	4961	5332	5449			
F. Groundwater	15541	16286	19776	100	281	389
G. Water and soils	9816	10047	10365			
H. Lakes	7561	10661	11412	800	4437	5270
I. Water in plants	10214	14487	12406			
J. Erosion and sedimentation	14765	14577	15343			
K. Chemical processes	9882	12762	10912			
L. Estuaries	3459	11404	13669		8596	10338
M. Global water cycle problems	12143	12024	12073			
Subtotal	*127026*	*150835*	*151147*	*900*	*13314*	*15997*
III. Water Supply Augmentation and Conservation						
A. Saline water conversion	4699	4439	3719			
B. Water yield improvement	3131	3784	4764			
C. Use of water of impaired quality	856	1353	1128			
D. Conservation in domestic and municipal use	110	7	20			
E. Conservation in industrial use	375	296	426			
F. Conservation in agricultural use	4224	4577	4796			
Subtotal	*13394*	*14456*	*14853*			
IV. Water Quantity Management and Control						
A. Control of water on surface	22391	24480	23333			
B. Groundwater management	4420	3227	5485			
C. Effects on water of man's nonwater activities	3353	4378	4526			
D. Watershed activities	11114	13544	13839	6716	8990	9421
Subtotal	*41278*	*45629*	*47183*	*6716*	*8990*	*9421*

NSF			USGS			USBR		
1999	*2000*	*2001*	*1999*	*2000*	*2001*	*1999*	*2000*	*2001*
680	1087	1251						
6305	9551	10474	614	515	534			
6985	*10638*	*11725*	*614*	*515*	*534*			
503	3059	662	1741	1741	1648			
7824	7643	7627				269	232	168
811	749	949	1113	1230	1133			
3068	3504	3221	705	686	747	103	361	293
291	357	369	2112	2039	1981	235	336	279
1534	1598	2077	10631	10977	13479	202	225	48
2454	2994	2837	1065	954	1078			
861	370	849	1854	1819	1907			
3223	6492	3763						
421	512	540	5148	4217	4314	223	551	638
3898	7187	5123	5574	4988	5478			
661	186	537	2382	2206	2228			
8012	7652	7563	1417	1043	1249			
33561	*42302*	*36118*	*33742*	*31900*	*35242*	*1032*	*1705*	*1426*
						4699	4439	3719
131	30	45				23	135	121
81		535						
106	2	14						
54		356						
372	*32*	*950*				*4722*	*4574*	*3840*
158	1126	170	381	391	399	2037	2712	2657
100	1115	95	436	433	452	192	26	159
292	835	190	194	260	316			
155	107	106						
705	*3183*	*561*	*1011*	*1084*	*1167*	*2229*	*2738*	*2816*

continued

TABLE B-1 Continued

Research Category	All Federal Agencies			EPA		
	1999	*2000*	*2001*	*1999*	*2000*	*2001*
V. Water Quality Management and Protection						
A. Identification of pollutants	28055	27433	34754	11226	9966	10323
B. Sources and fate of pollution	63604	61622	67469	3785	4497	4713
C. Effects of pollution on water resources	33152	30999	31661	6541	5399	4265
D. Waste treatment processes	23535	23845	26344			
E. Ultimate disposal of wastes	4567	4753	4845			
F. Water treatment and distribution	8761	11623	11692	6699	9182	9273
G. Water quality control	17677	18922	25919	3213	5887	4826
H. Effects of waterborne pollution on human health	10961	12472	17777	4628	2212	6200
Subtotal	*190311*	*191669*	*220460*	*36092*	*37143*	*39600*
VI. Water Resources Planning and Other Institutional Issues						
A. Techniques of planning	2286	2194	991			
B. Evaluation process	1998	1729	470	455	439	
C. Economic issues	1687	881	935			
D. Water demand	641	746	593			
E. Water law and institutions	1035	40	45			
F. Nonstructural alternatives	250	286	311			
G. Risk perception and communication	1526	2583	1061	331	465	110
H. Other social sciences	88	233	280			
I. Water resources policy	1360	1141	391			
Subtotal	*10871*	*9834*	*5077*	*786*	*904*	*110*
VII. Resources Data						
A. Network design	1866	1990	1990	295	262	
B. Data acquisition	3750	4033	4962			750
C. Evaluation, processing, and publication	2693	2656	2335			
Subtotal	*8309*	*8679*	*9286*	*295*	*262*	*750*

NSF			USGS			USBR		
1999	*2000*	*2001*	*1999*	*2000*	*2001*	*1999*	*2000*	*2001*
3795	5228	4945	3884	3753	4392	207	202	251
9935	12267	12501	13571	13460	12977	169	296	164
1539	784	984	13869	13377	14295			
3333	4639	5311				646	845	895
845	1021	984	1758	1698	1762			
1428	1988	2276						
1805	1182	1168	146	95	333	772	825	668
765	1128	1124						
23444	*28237*	*29292*	*33228*	*32383*	*33759*	*1794*	*2168*	*1978*
1023	762	30						
1023	762	30						
1206	333	314				36	98	166
206	333	314	150	138				
1000								
						25	60	84
75	845	109				42	14	
	136	240						
1149	930	203				49	60	168
5682	*4102*	*1240*	*150*	*138*		*152*	*232*	*418*
84	137	146	400	467	619			
84	137	146	1602	1622	1856	79	145	40
72	117	125				1	40	55
240	*391*	*417*	*2002*	*2089*	*2475*	*80*	*185*	*95*

continued

TABLE B-1 Continued

Research Category	All Federal Agencies 1999	All Federal Agencies 2000	All Federal Agencies 2001	EPA 1999	EPA 2000	EPA 2001
VIII. Engineering Works						
A. Structures	8912	10321	8918			
B. Hydraulics	106	224	142			
C. Hydraulic machinery	574	882	735			
D. Soil mechanics	1619	1130	1982			
E. Rock mechanics and geology						
F. Concrete	576	340	483			
G. Materials	1570	1671	1232			
H. Rapid excavation						
I. Fisheries engineering	20143	39757	32469			
J. Infrastructure repair and rehabilitation	3303	3734	2792	230		
K. Restoration engineering						
L. Facility protection	150	59	199			
Subtotal	*36953*	*58118*	*48951*	*230*		
IX. Manpower, Grants, and Facilities						
A. Education—extramural	2187	4973	7925			
B. Education—in-house	22	32	31			
C. Capital expenditures for research facilities	2060	19171	17873		187	
D. Grants, contracts, and research act allotments	3818	3818	4218			
Subtotal	*8087*	*27994*	*30047*		*187*	
X. Scientific and Technical Information						
A. Acquisition and processing	15	15	15			
B. Reference and retrieval	15	15	15			
C. Secondary publication and distribution			135			
D. Specialized information center services	751	824	1332			
E. Translations						
F. Preparation of reviews	179	315	352			
Subtotal	*960*	*1168*	*1849*			
XI. Aquatic Ecosystem Management and Protection						
A. Ecosystem and habitat conservation	59932	61813	63639	12472	10016	12242
B. Aquatic ecosystem assessment	59940	62139	60619	31869	28056	27082
C. Effects of climate change	13446	9593	16611	3069		5328
D. Biogeochemical cycles	21331	24891	22485	147	98	124
Subtotal	*154648*	*158436*	*163354*	*47557*	*38170*	*44776*
Total Water Resources Research	**599436**	**677971**	**705336**	**92576**	**98970**	**110654**

NSF			USGS			USBR		
1999	*2000*	*2001*	*1999*	*2000*	*2001*	*1999*	*2000*	*2001*
						106	224	142
						574	882	735
716	384	1292				96	116	133
						49	52	95
						253	244	182
			2000	2000	2000	443	457	769
275	148	497				117	113	228
150	59	199						
1141	*591*	*1988*	*2000*	*2000*	*2000*	*1638*	*2088*	*2284*
1337	3049	6112						
2060	4860	3604		14124	14269			
			3818	3818	4218			
3397	*7909*	*9716*	*3818*	*17942*	*18487*			
		135						
751	824	1332						
157	293	330						
908	*1116*	*1797*						
10260	13006	11274	15837	15486	15970	270	201	354
10791	15468	13552	13360	14336	14561	254	316	444
5454	5848	6354	2624	2395	2368			
14365	18069	14383	2840	2840	2808			
40869	*52391*	*45563*	*34661*	*35057*	*35707*	*524*	*517*	*798*
117303	**150892**	**139365**	**111226**	**123108**	**129371**	**12171**	**14207**	**13655**

TABLE B-2 Funding Levels for Water Resources Research for the USDA for FY1999–2001 (data in the thousands of dollars; not adjusted for inflation)

	USDA Total			ARS		
Research Category	*1999*	*2000*	*2001*	*1999*	*2000*	*2001*
II. Water Cycle						
A. General	2919	2933	2495	2919	2668	2330
B. Precipitation	2363	2658	2446	2363	2658	2446
C. Snow, ice, and frost	971	997	1099	971	997	1099
D. Evaporation and transpiration	3889	4009	4401	3889	4009	4401
E. Streamflow and runoff	1196	1465	1674	1196	1465	1494
F. Groundwater	2754	2880	3453	2754	2880	3453
G. Water and soils	4847	4649	5000	4576	4399	4845
H. Lakes	644	650	0	644	650	
I. Water in plants	6441	7445	8093	6441	7445	8093
J. Erosion and sedimentation	8596	8797	9351	6672	6880	7121
K. Chemical processes	160	337	61	60	60	61
L. Estuaries	16	16	166	16	16	166
M. Global water cycle problems	1474	2089	2021	1474	2089	2021
Subtotal	*36270*	*38925*	*40260*	*33975*	*36216*	*37530*
III. Water Supply Augmentation and Conservation						
A. Saline water conversion						
B. Water yield improvement	2977	3619	4598	2977	3619	4598
C. Use of water of impaired quality	775	953	963	775	953	963
D. Conservation in domestic and municipal use						
E. Conservation in industrial use	318	292	65	210	210	
F. Conservation in agricultural use	4218	4570	4788	4054	4344	4684
Subtotal	*8288*	*9434*	*10414*	*8016*	*9126*	*10245*
IV. Water Quantity Management and Control						
A. Control of water on surface	10805	11108	11198	1275	809	798
B. Groundwater management	3672	1628	4749	3672	1628	4598
C. Effects on water of man's nonwater activities	2867	3283	4020			
D. Watershed activities	4243	4447	4312	3963	4447	4312
Subtotal	*21587*	*20466*	*24279*	*8910*	*6884*	*9708*

CSREES			ERS			USFS		
1999	*2000*	*2001*	*1999*	*2000*	*2001*	*1999*	*2000*	*2001*
	265	165						
		180						
271	250	155						
						1924	1917	2230
100	277							
371	*792*	*500*				*1924*	*1917*	*2230*
108	82	65						
164	226	104						
272	*308*	*169*						
8466	9221	9254				1064	1078	1146
		151						
		290				2867	3283	3730
280								
8746	*9221*	*9695*				*3931*	*4361*	*4876*

continued

TABLE B-2 Continued

	USDA Total			ARS		
Research Category	*1999*	*2000*	*2001*	*1999*	*2000*	*2001*
V. Water Quality Management and Protection						
A. Identification of pollutants	1112	1007	1013	365	466	155
B. Sources and fate of pollution	7394	8710	10290	6213	7502	8814
C. Effects of pollution on water resources	980	1851	2162	914	1206	1565
D. Waste treatment processes	1408	1467	2411	639	781	538
E. Ultimate disposal of wastes	1964	2034	2099	1964	2034	2099
F. Water treatment and distribution	634	453	143	634	453	143
G. Water quality control	11726	10915	18903	6915	6072	9106
H. Effects of waterborne pollution on human health						
Subtotal	*25218*	*26437*	*37021*	*17644*	*18514*	*22420*
VI. Water Resources Planning and Other Institutional Issues						
A. Techniques of planning	53	404	54	53	53	54
B. Evaluation process	275	275	275			
C. Economic issues	420	420	420			
D. Water demand	270	257	258	92	92	93
E. Water law and institutions						
F. Nonstructural alternatives	220	220	220			
G. Risk perception and communication		260				
H. Other social sciences						
I. Water resources policy						
Subtotal	*1238*	*1836*	*1227*	*145*	*145*	*147*
VII. Resources Data						
A. Network design	712	744	780	712	744	780
B. Data acquisition	88	89	90	88	89	90
C. Evaluation, processing, and publication	283	178	180	283	178	180
Subtotal	*1083*	*1011*	*1050*	*1083*	*1011*	*1050*
IX. Manpower, Grants, and Facilities						
A. Education—extramural		974	554			
B. Education—in-house						
C. Capital expenditures for research facilities						
D. Grants, contracts, and research act allotments						
Subtotal		*974*	*554*			

CSREES			ERS			USFS		
1999	*2000*	*2001*	*1999*	*2000*	*2001*	*1999*	*2000*	*2001*
747	541	858						
1181	1208	1476						
66	645	597						
769	686	1873						
4811	4843	9797						
7574	*7923*	*14601*						
	351							
			275	275	275			
			220	220	220	200	200	200
			178	165	165			
			220	220	220			
	260							
	611		*893*	*880*	*880*	*200*	*200*	*200*
	974	554						
	974	*554*						

continued

TABLE B-2 Continued

Research Category	USDA Total			ARS		
	1999	*2000*	*2001*	*1999*	*2000*	*2001*
XI. Aquatic Ecosystem Management and Protection						
A. Ecosystem and habitat conservation	11923	13786	14845			
B. Aquatic ecosystem assessment	110	110	110			
C. Effects of climate change						
D. Biogeochemical cycles	3203	3147	4289			
Subtotal	*15236*	*17043*	*19244*			
Total Water Resources Research	**108920**	**116126**	**134049**	**69773**	**71896**	**81100**

CSREES			ERS			USFS		
1999	*2000*	*2001*	*1999*	*2000*	*2001*	*1999*	*2000*	*2001*
1080	2119	2244				10843	11667	12601
			110	110	110			
	89	315				3203	3058	3974
1080	*2208*	*2559*	*110*	*110*	*110*	*14046*	*14725*	*16575*
18043	**22037**	**28078**	**1003**	**990**	**990**	**20101**	**21203**	**23881**

TABLE B-3 Funding Levels for Water Resources Research for DoD and DOE for FY1999–2001 (data in the thousands of dollars; not adjusted for inflation)

	DoD Total			Corps		
Research Category	*1999*	*2000*	*2001*	*1999*	*2000*	*2001*
II. Water Cycle						
A. General	3098	4348	3790	3098	4348	3790
B. Precipitation						
C. Snow, ice, and frost	950	719	666	950	719	666
D. Evaporation and transpiration						
E. Streamflow and runoff						
F. Groundwater						
G. Water and soils						
H. Lakes						
I. Water in plants						
J. Erosion and sedimentation	377	500	500	377	500	500
K. Chemical processes						
L. Estuaries						
M. Global water cycle problems						
Subtotal	*4425*	*5567*	*4956*	*4425*	*5567*	*4956*
III. Water Supply Augmentation and Conservation						
A. Saline water conversion						
B. Water yield improvement						
C. Use of water of impaired quality						
D. Conservation in domestic and municipal use						
E. Conservation in industrial use						
F. Conservation in agricultural use						
Subtotal						
IV. Water Quantity Management and Control						
A. Control of water on surface	8950	8867	8724	8950	8867	8724
B. Groundwater management						
C. Effects on water of man's nonwater activities						
D. Watershed activities						
Subtotal	*8950*	*8867*	*8724*	*8950*	*8867*	*8724*

ONR			SERDP			DOE		
1999	*2000*	*2001*	*1999*	*2000*	*2001*	*1999*	*2000*	*2001*
							400	500
							400	*500*
							203	100
							203	*100*

continued

TABLE B-3 Continued

Research Category	DoD Total			Corps		
	1999	*2000*	*2001*	*1999*	*2000*	*2001*
V. Water Quality Management and Protection						
A. Identification of pollutants	2851	2977	2730			
B. Sources and fate of pollution	2500	3743	3388			
C. Effects of pollution on water resources	7994	7455	7462	4000	5031	4939
D. Waste treatment processes	10298	9944	10427			
E. Ultimate disposal of wastes						
F. Water treatment and distribution						
G. Water quality control						
H. Effects of waterborne pollution on human health						
Subtotal	*23643*	*24119*	*24007*	*4000*	*5031*	*4939*
VI. Water Resources Planning and Other Institutional Issues						
A. Techniques of planning	782	692	680	782	692	680
B. Evaluation process						
C. Economic issues						
D. Water demand						
E. Water law and institutions						
F. Nonstructural alternatives						
G. Risk perception and communication	1068	981	817	1068	981	817
H. Other social sciences						
I. Water resources policy						
Subtotal	*1850*	*1673*	*1497*	*1850*	*1673*	*1497*
VII. Resources Data						
A. Network design						
B. Data acquisition	1162	1312	1314	1162	1312	1314
C. Evaluation, processing, and publication	1679	1659	1242	1679	1659	1242
Subtotal	*2841*	*2971*	*2556*	*2841*	*2971*	*2556*

ONR			SERDP			DOE		
1999	*2000*	*2001*	*1999*	*2000*	*2001*	*1999*	*2000*	*2001*
			2851	2977	2730	2100	1200	1300
2500	2754	2218		989	1170	24900	17300	21700
			3994	2424	2523			
			10298	9944	10427	7850	6950	7300
2500	*2754*	*2218*	*17143*	*16334*	*16850*	*34850*	*25450*	*30300*

continued

TABLE B-3 Continued

Research Category	DoD Total *1999*	DoD Total *2000*	DoD Total *2001*	Corps *1999*	Corps *2000*	Corps *2001*
VIII. Engineering Works						
A. Structures	8912	10321	8918	8912	10321	8918
B. Hydraulics						
C. Hydraulic machinery						
D. Soil mechanics	807	630	557	807	630	557
E. Rock mechanics and geology						
F. Concrete	527	288	388	527	288	388
G. Materials	1317	1427	1050	1317	1427	1050
H. Rapid excavation						
I. Fisheries engineering	17700	37300	29700	17700	37300	29700
J. Infrastructure repair and rehabilitation	2681	3473	2067	2681	3473	2067
K. Restoration engineering						
L. Facility protection						
Subtotal	*31944*	*53439*	*42680*	*31944*	*53439*	*42680*
X. Scientific and Technical Information						
A. Acquisition and processing	15	15	15			
B. Reference and retrieval	15	15	15			
C. Secondary publication and distribution						
D. Specialized information center services						
E. Translations						
F. Preparation of reviews	22	22	22			
Subtotal	*52*	*52*	*52*			
XI. Aquatic Ecosystem Management and Protection						
A. Ecosystem and habitat conservation	6759	6385	6078	6759	6385	6078
B. Aquatic ecosystem assessment	1630	1595	1961	1630	1595	1961
C. Effects of climate change						
D. Biogeochemical cycles						
Subtotal	*8389*	*7980*	*8039*	*8389*	*7980*	*8039*
Total Water Resources Research	**82094**	**104668**	**92511**	**62399**	**85528**	**73391**

ONR			SERDP			DOE		
1999	*2000*	*2001*	*1999*	*2000*	*2001*	*1999*	*2000*	*2001*
			15	15	15			
			15	15	15			
			22	22	22			
			52	*52*	*52*			
2500	**2754**	**2218**	**17195**	**16386**	**16902**	**34850**	**26053**	**30900**

TABLE B-4 Funding Levels for Water Resources Research for NOAA and NASA for FY1999–2001 (data in the thousands of dollars; not adjusted for inflation)

	NOAA Total			NWS		
Research Category	*1999*	*2000*	*2001*	*1999*	*2000*	*2001*
II. Water Cycle						
A. General	1374	1128	695	505	523	566
B. Precipitation	1913	2160	2590	530	550	572
C. Snow, ice, and frost	745	769	775	150	169	170
D. Evaporation and transpiration	575	580	586	175	180	186
E. Streamflow and runoff	377	385	396	277	285	296
F. Groundwater	20	25	30			
G. Water and soils	150	150	150			
H. Lakes	3152	3135	3136			
I. Water in plants	150	150	150			
J. Erosion and sedimentation						
K. Chemical processes						
L. Estuaries						
M. Global water cycle problems	240	240	240			
Subtotal	*8696*	*8722*	*8748*	*1637*	*1707*	*1790*
III. Water Supply Augmentation and Conservation						
A. Saline water conversion						
B. Water yield improvement						
C. Use of water of impaired quality						
D. Conservation in domestic and municipal use	4	5	6			
E. Conservation in industrial use	3	4	5			
F. Conservation in agricultural use	6	7	8			
Subtotal	*13*	*16*	*19*			
IV. Water Quantity Management and Control						
A. Control of water on surface	60	73	85			
B. Groundwater management	20	25	30			
C. Effects on water of man's nonwater activities						
D. Watershed activities						
Subtotal	*80*	*98*	*115*			

OOAR			NOS			NASA		
1999	*2000*	*2001*	*1999*	*2000*	*2001*	*1999*	*2000*	*2001*
869	605	129				1000	1000	1000
1383	1610	2018				2000	2000	2000
595	600	605				500	500	500
400	400	400				250	250	250
100	100	100				750	750	750
20	25	30				300	300	300
150	150	150				1300	1300	1300
3152	3135	3136				250	250	250
150	150	150				400	400	400
						250	250	250
						400	400	400
240	240	240				1000	1000	1000
7059	*7015*	*6958*				*8400*	*8400*	*8400*
4	5	6						
3	4	5						
6	7	8						
13	*16*	*19*						
60	73	85						
20	25	30						
80	*98*	*115*						

continued

TABLE B-4 Continued

Research Category	NOAA Total 1999	NOAA Total 2000	NOAA Total 2001	NWS 1999	NWS 2000	NWS 2001
V. Water Quality Management and Protection						
A. Identification of pollutants	2880	3100	9800			
B. Sources and fate of pollution	1350	1349	1736			
C. Effects of pollution on water resources	2229	2133	2493			
D. Waste treatment processes						
E. Ultimate disposal of wastes						
F. Water treatment and distribution						
G. Water quality control	15	18	21			
H. Effects of waterborne pollution on human health						
Subtotal	*6474*	*6600*	*14050*			
VI. Water Resources Planning and Other Institutional Issues						
A. Techniques of planning	428	336	227			
B. Evaluation process	245	253	165			
C. Economic issues	25	30	35			
D. Water demand	15	18	21			
E. Water law and institutions	35	40	45			
F. Nonstructural alternatives	5	6	7			
G. Risk perception and communication	10	18	25			
H. Other social sciences	88	97	40			
I. Water resources policy	162	151	20			
Subtotal	*1013*	*949*	*585*			
VII. Resources Data						
A. Network design	125	130	195	120	120	180
B. Data acquisition	235	228	266	160	160	160
C. Evaluation, processing, and publication	408	412	483	200	200	265
Subtotal	*768*	*770*	*944*	*480*	*480*	*605*
IX. Manpower, Grants, and Facilities						
A. Education—extramural	850	950	1259	50	150	200
B. Education—in-house	22	32	31	22	32	31
C. Capital expenditures for research facilities						
D. Grants, contracts, and research act allotments						
Subtotal	*872*	*982*	*1290*	*72*	*182*	*231*

OOAR			NOS			NASA		
1999	*2000*	*2001*	*1999*	*2000*	*2001*	*1999*	*2000*	*2001*
			2880	3100	9800			
			1350	1349	1736			
729	915	927	1500	1218	1566			
15	18	21						
744	*933*	*948*	*5730*	*5667*	*13102*			
428	336	227						
245	253	165						
25	30	35						
15	18	21						
35	40	45						
5	6	7						
10	18	25						
88	97	40						
162	151	20						
1013	*949*	*585*						
5	10	15				250	250	250
5	12	20	70	56	86	500	500	500
208	212	218				250	250	250
218	*234*	*253*	*70*	*56*	*86*	*1000*	*1000*	*1000*
			800	800	1059			
			800	*800*	*1059*			

continued

TABLE B-4 Continued

	NOAA Total			NWS		
Research Category	*1999*	*2000*	*2001*	*1999*	*2000*	*2001*
XI. Aquatic Ecosystem Management and Protection						
A. Ecosystem and habitat conservation	2411	2933	2876			
B. Aquatic ecosystem assessment	1926	2258	2909			
C. Effects of climate change	1899	950	2161			
D. Biogeochemical cycles	476	437	581			
Subtotal	*6712*	*6578*	*8527*			
Total Water Resources Research	**24628**	**24715**	**34278**	**2189**	**2369**	**2626**

OOAR			NOS			NASA		
1999	*2000*	*2001*	*1999*	*2000*	*2001*	*1999*	*2000*	*2001*
91	558	476	2320	2375	2400			
536	875	739	1390	1383	2170			
929	850	1581	970	100	580	400	400	400
476	437	581				300	300	300
2032	*2720*	*3377*	*4680*	*3858*	*5150*	*700*	*700*	*700*
11159	**11965**	**12255**	**11280**	**10381**	**19397**	**10100**	**10100**	**10100**

TABLE B-5 Funding Levels for Water Resources Research for DHHS for FY1999–2001 (data in the thousands of dollars; not adjusted for inflation)

	DHHS Total		
Research Category	*1999*	*2000*	*2001*
V. Water Quality Management and Protection			
A. Identification of pollutants			
B. Sources and fate of pollution			
C. Effects of pollution on water resources			
D. Waste treatment processes			
E. Ultimate disposal of wastes			
F. Water treatment and distribution			
G. Water quality control			
H. Effects of waterborne pollution on human health	5568	9132	10453
Subtotal	*5568*	*9132*	*10453*
Total Water Resources Research	**5568**	**9132**	**10453**

NIEHS			ATSDR			NCI		
1999	*2000*	*2001*	*1999*	*2000*	*2001*	*1999*	*2000*	*2001*
2671	6128	7109	2837	2404	1844	60	600	1500
2671	*6128*	*7109*	*2837*	*2404*	*1844*	*60*	*600*	*1500*
2671	**6128**	**7109**	**2837**	**2404**	**1844**	**60**	**600**	**1500**

TABLE B-6 Funding Levels for Water Resources Research for All the Nonfederal Organizations Queried, and Levels for WERF, AWWARF, and TNC for FY1999–2001 (data in the thousands of dollars; not adjusted for inflation)

	Nonprofit Total		
Research Category	*1999*	*2000*	*2001*
I. Nature of Water			
A. Properties of water			25
B. Aqueous solutions			49
Subtotal			*74*
II. Water Cycle			
A. General	222	118	256
B. Precipitation	23	28	59
C. Snow, ice, and frost	32	34	61
D. Evaporation and transpiration	6	9	60
E. Streamflow and runoff	211	36	372
F. Groundwater	92	99	569
G. Water and soils	6	6	18
H. Lakes			15
I. Water in plants			20
J. Erosion and sedimentation	320	457	564
K. Chemical processes	220	269	1390
L. Estuaries			21
M. Global water cycle problems			25
Subtotal	*1132*	*1056*	*3430*
III. Water Supply Augmentation and Conservation			
A. Saline water conversion			45
B. Water yield improvement	736	581	593
C. Use of water of impaired quality	470	423	353
D. Conservation in domestic and municipal use		204	560
E. Conservation in industrial use			
F. Conservation in agricultural use		772	978
Subtotal	*1206*	*1980*	*2529*
IV. Water Quantity Management and Control			
A. Control of water on surface	350	184	77
B. Groundwater management	272	22	121
C. Effects on water of man's nonwater activities			10
D. Watershed activities			8
Subtotal	*622*	*206*	*216*

WERF			AWWARF			TNC		
1999	*2000*	*2001*	*1999*	*2000*	*2001*	*1999*	*2000*	*2001*
200		100						
					350			
200		*100*			*350*			
			470	75				
					350			
			470	*75*	*350*			
200	100							
			250					
200	*100*		*250*					

continued

TABLE B-6 Continued

Research Category	Nonprofit Total		
	1999	*2000*	*2001*
V. Water Quality Management and Protection			
A. Identification of pollutants	1561	2482	3495
B. Sources and fate of pollution	1438	1449	1275
C. Effects of pollution on water resources	3189	1222	1152
D. Waste treatment processes	2589	3307	3384
E. Ultimate disposal of wastes	1230	850	604
F. Water treatment and distribution	2553	1910	3747
G. Water quality control	1198	1943	3169
H. Effects of waterborne pollution on human health	1509	1289	754
Subtotal	*15267*	*14452*	*17580*
VI. Water Resources Planning and Other Institutional Issues			
A. Techniques of planning	1021	834	1580
B. Evaluation process	514	313	200
C. Economic issues	450	180	168
D. Water demand	103	52	181
E. Water law and institutions		333	33
F. Nonstructural alternatives	399	402	331
G. Risk perception and communication	1317	1014	1261
H. Other social sciences			455
I. Water resources policy	13	13	15
Subtotal	*3817*	*3141*	*4224*
VII. Resources Data			
A. Network design	28	76	64
B. Data acquisition	260	462	126
C. Evaluation, processing, and publication	25	289	455
Subtotal	*313*	*827*	*645*
VIII. Engineering Works			
A. Structures	317	429	504
B. Hydraulics	892	923	1089
C. Hydraulic machinery		100	60
D. Soil mechanics		2	33
E. Rock mechanics and geology			8
F. Concrete	100		
G. Materials			
H. Rapid excavation			
I. Fisheries engineering	15		
J. Infrastructure repair and rehabilitation	550	1142	913
K. Restoration engineering	5		
L. Facility protection		200	150
Subtotal	*1879*	*2796*	*2757*

WERF			AWWARF			TNC		
1999	*2000*	*2001*	*1999*	*2000*	*2001*	*1999*	*2000*	*2001*
300	700	1700	861	1360	1315			
300		100	338	450	203			
2300	800	700	350					
2000	2500	1900						
1000	500	400	230	350	150			
		200	2346	1760	3360			
100	500	800		200	750			
	500	300	1300	500	75			
6000	*5500*	*6100*	*5425*	*4620*	*5853*			
			840		650			
300		200	175	300				
			450		150			
				300				
	300		700		500			
					455			
300	*300*	*200*	*2165*	*600*	*1755*			
				200				
	200				375			
	200			*200*	*375*			
		200						
	100							
100								
100		200	450	650	250			
				200	150			
200	*100*	*400*	*450*	*850*	*400*			

continued

TABLE B-6 Continued

	Nonprofit Total		
Research Category	*1999*	*2000*	*2001*
IX. Manpower, Grants, and Facilities			
A. Education—extramural	2217	3060	3788
B. Education—in-house	188		
C. Capital expenditures for research facilities			
D. Grants, contracts, and research act allotments			
Subtotal	*2405*	*3060*	*3788*
X. Scientific and Technical Information			
A. Acquisition and processing	20	20	40
B. Reference and retrieval	20	2	2
C. Secondary publication and distribution			
D. Specialized information center services			
E. Translations			
F. Preparation of reviews		37	3
Subtotal	*40*	*59*	*45*
XI. Aquatic Ecosystem Management and Protection			
A. Ecosystem and habitat conservation	9161	9905	9614
B. Aquatic ecosystem assessment	903	1148	1286
C. Effects of climate change			15
D. Biogeochemical cycles			100
Subtotal	*10064*	*11053*	*11015*
Total Water Resources Research	**36745**	**38630**	**46303**

WERF			AWWARF			TNC		
1999	*2000*	*2001*	*1999*	*2000*	*2001*	*1999*	*2000*	*2001*
	500					9050	9300	9500
	500					*9050*	*9300*	*9500*
6900	**6700**	**6800**	**8760**	**6345**	**9083**	**9050**	**9300**	**9500**

TABLE B-7 Funding Levels for Water Resources Research for the four largest WRRIs for FY1999–2001 (data in the thousands of dollars; not adjusted for inflation)

	Nevada		
Research Category	*1999*	*2000*	*2001*
I. Nature of Water			
A. Properties of water			
B. Aqueous solutions			
Subtotal			
II. Water Cycle			
A. General	14		
B. Precipitation			
C. Snow, ice, and frost			
D. Evaporation and transpiration			
E. Streamflow and runoff			29
F. Groundwater			
G. Water and soils			
H. Lakes			
I. Water in plants			
J. Erosion and sedimentation			
K. Chemical processes			
L. Estuaries			
M. Global water cycle problems			
Subtotal	*14*		*29*
III. Water Supply Augmentation and Conservation			
A. Saline water conversion			
B. Water yield improvement	136	37	
C. Use of water of impaired quality			
D. Conservation in domestic and municipal use			
E. Conservation in industrial use			
F. Conservation in agricultural use			
Subtotal	*136*	*37*	
IV. Water Quantity Management and Control			
A. Control of water on surface		9	12
B. Groundwater management		6	61
C. Effects on water of man's nonwater activities			
D. Watershed activities			
Subtotal		*15*	*73*

Texas			Utah			Pennsylvania		
1999	*2000*	*2001*	*1999*	*2000*	*2001*	*1999*	*2000*	*2001*
		25						
		25						24
		50						*24*
		20	118	118	179	90		57
		18	23	28	41			
			32	34	61			
		50	6	9	10			
		40	11	16	31		20	172
		45	92	99	174			
		8	6	6	10			
		10			5			
		20						
		15				320	457	549
		18				220	269	1372
		21						
		25						
		290	*288*	*310*	*511*	*630*	*746*	*2150*
		45						
600	502	464					42	129
	348	353						
	204	210						
	772	978						
600	*1826*	*2050*					*42*	*129*
150	75	65						
		25	22	16	35			
		10						
		8						
150	*75*	*108*	*22*	*16*	*35*			

continued

TABLE B-7 Continued

	Nevada		
Research Category	*1999*	*2000*	*2001*
V. Water Quality Management and Protection			
A. Identification of pollutants			12
B. Sources and fate of pollution	77	73	124
C. Effects of pollution on water resources	117		
D. Waste treatment processes			
E. Ultimate disposal of wastes			
F. Water treatment and distribution			
G. Water quality control	133	50	144
H. Effects of waterborne pollution on human health			
Subtotal	*327*	*123*	*280*
VI. Water Resources Planning and Other Institutional Issues			
A. Techniques of planning		29	
B. Evaluation process			
C. Economic issues			
D. Water demand			
E. Water law and institutions			
F. Nonstructural alternatives			
G. Risk perception and communication			
H. Other social sciences			
I. Water resources policy			
Subtotal		*29*	
VII. Resources Data			
A. Network design			
B. Data acquisition			
C. Evaluation, processing, and publication			
Subtotal			
VIII. Engineering Works			
A. Structures			
B. Hydraulics			
C. Hydraulic machinery			
D. Soil mechanics		2	13
E. Rock mechanics and geology			
F. Concrete			
G. Materials			
H. Rapid excavation			
I. Fisheries engineering			
J. Infrastructure repair and rehabilitation			
K. Restoration engineering			
L. Facility protection			
Subtotal		*2*	*13*

Texas			Utah			Pennsylvania		
1999	*2000*	*2001*	*1999*	*2000*	*2001*	*1999*	*2000*	*2001*
		40	400	422	428			
		15	419	533	668	304	393	165
		30				422	422	422
		20	76	67	53	513	740	1411
								54
		10	57		27	150	150	150
		16	743	911	1123	222	282	336
		5	209	289	374			
		136	*1904*	*2222*	*2673*	*1611*	*1987*	*2538*
		120	116	104	86	65	701	724
			39	13				
		18					180	
		120	103	52	61			
	33	33						
			399	402	331			
		5	617	714	748			8
		15	13	13				
	33	*311*	*1287*	*1298*	*1226*	*65*	*881*	*732*
			28	76	64			
			37	106	113	223	156	13
			25	89	80			
			90	*271*	*257*	*223*	*156*	*13*
		40	113	126	66	204	303	198
		60	892	923	1029			
		60						
		20						
		8						
			15					
	492	463						
			5					
	492	*651*	*1025*	*1049*	*1095*	*204*	*303*	*198*

TABLE B-7 Continued

Research Category	Nevada		
	1999	*2000*	*2001*
IX. Manpower, Grants, and Facilities			
A. Education—extramural			
B. Education—in-house			
C. Capital expenditures for research facilities			
D. Grants, contracts, and research act allotments			
Subtotal			
X. Scientific and Technical Information			
A. Acquisition and processing			
B. Reference and retrieval	20		
C. Secondary publication and distribution			
D. Specialized information center services			
E. Translations			
F. Preparation of reviews		37	3
Subtotal	*20*	*37*	*3*
XI. Aquatic Ecosystem Management and Protection			
A. Ecosystem and habitat conservation			
B. Aquatic ecosystem assessment			
C. Effects of climate change			
D. Biogeochemical cycles			
Subtotal			
Total Water Resources Research	**497**	**243**	**398**

Texas			Utah			Pennsylvania		
1999	*2000*	*2001*	*1999*	*2000*	*2001*	*1999*	*2000*	*2001*
100	1010	1052	422	588	619	1695	1462	2117
						188		
100	*1010*	*1052*	*422*	*588*	*619*	*1883*	*1462*	*2117*
20	20	40						
				2	2			
20	*20*	*40*		*2*	*2*			
		40	94	88	57	17	17	17
		20	759	1009	1092	144	139	174
		15						100
		75	*853*	*1097*	*1149*	*161*	*156*	*291*
870	**3456**	**4763**	**5891**	**6853**	**7567**	**4777**	**5733**	**8192**

Appendix C

Likelihood of Differences in U.S. Water Resources Research Funding Levels Between the Mid 1970s and the Late 1990s

ANALYSIS GOALS AND METHODS

As stated in Chapter 4, the historical research funding data and those collected as part of the present assessment effort contain significant uncertainty. It is thus appropriate to cast any comparison between funding levels in different eras in a likelihood framework. The specific question addressed in this appendix is: *How likely is it that the level of water resources research funding in the late 1990s is greater or less than the analogous level of the mid 1970s?*

This question was examined for the total water resources research funding and for the funding in each major Federal Coordinating Council of Science, Engineering, and Technology (FCCSET) category under a number of assumptions, considered as most reasonable by the committee. These assumptions regard the character of data uncertainty. In particular, it was assumed that the annual data contain measurement errors that are independent from year to year, that the distribution of errors in averages of annual values can be well approximated by a normal distribution, that the standard deviation of the errors in averages of annual values ranges in all cases from 25 percent of the average to 50 percent of the average, and that there are no significant systematic biases in the annual funding data.

The available funding data consist of annual funding estimates as a total and by major FCCSET category for 1973, 1974, and 1975 (early period) and 1999, 2000, and 2001 (recent period). In all cases of analysis, averages of the annual values in the early period and averages of the annual values in the recent period were obtained. A total of 10,000 samples were generated from the assumed normal distributions that have as means the respective averages of the early and recent periods and as standard deviations values that are first 25 percent (best guess) and then 50 percent (high value) of the respective mean. The differences between the members of each pair (the mid 1970s average minus the late 1990s average) were computed in each funding case, and the cumulative histograms of the differences were plotted on normal probability paper. The cumulative probability value for which the difference was less than zero (1970s funding minus 1990s funding < 0) was also computed and designated in the cumulative distribution plots.

The cumulative probability plots and information on them may be used to answer the posed question. If the zero-level cumulative probability value mentioned above is near 50 percent (say within 10 percent up or down), then it is doubtful that differences in funding levels are significant. If it is significantly lower than 50 percent then there is high likelihood that the late 1990s funding level is indeed lower than the mid 1970s level. If it is significantly greater than 50 percent then there is high likelihood that the late 1990s funding level is higher than the mid 1970s level. Results and significant findings for total water resources funding and for funding in each major FCCSET category are presented below.

TOTAL WATER RESOURCES RESEARCH

It is evident from Figure C-1 that the likelihood that the means are different under the stated conditions of uncertainty is small, since the values on either side of 0 (zero) are close to 50 percent. This makes zero the median and mean of the distribution of differences.

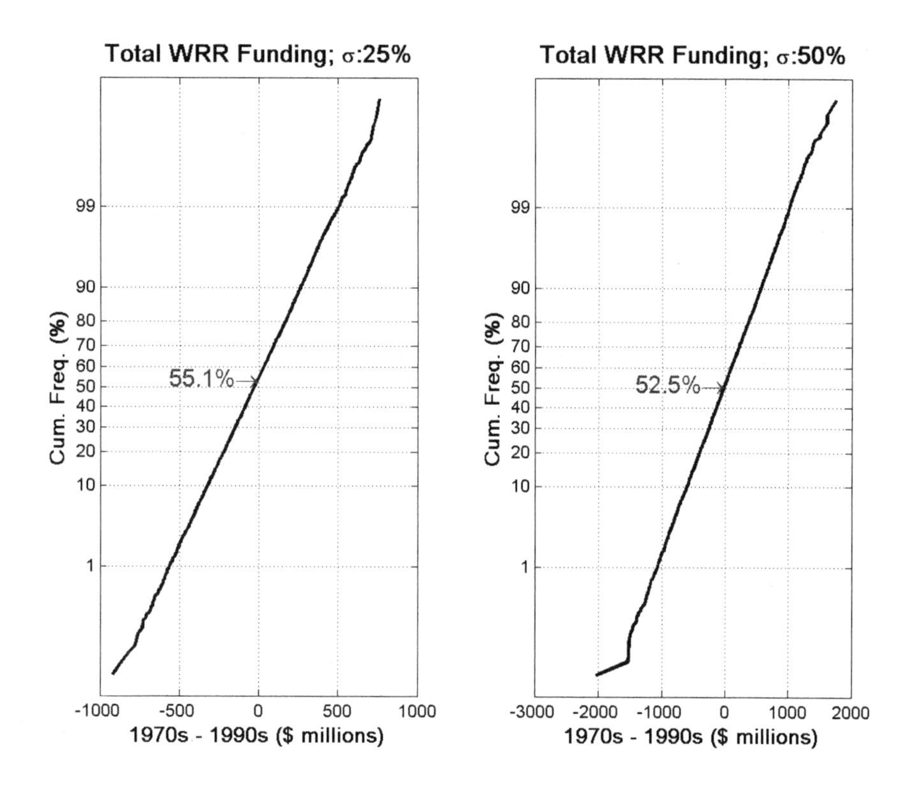

FIGURE C-1 Cumulative frequency distributions of 10,000 samples of generated differences between means of years 1973, 1974, 1975 and of years 1999, 2000, 2001 for total water resources research (WRR) funding.

CATEGORY I (NATURE OF WATER)

As shown in Figure C-2, it is very clear that the likelihood that the late 1990s funding is greater than the mid 1970s funding is very high, independent of the level of standard deviation assumed.

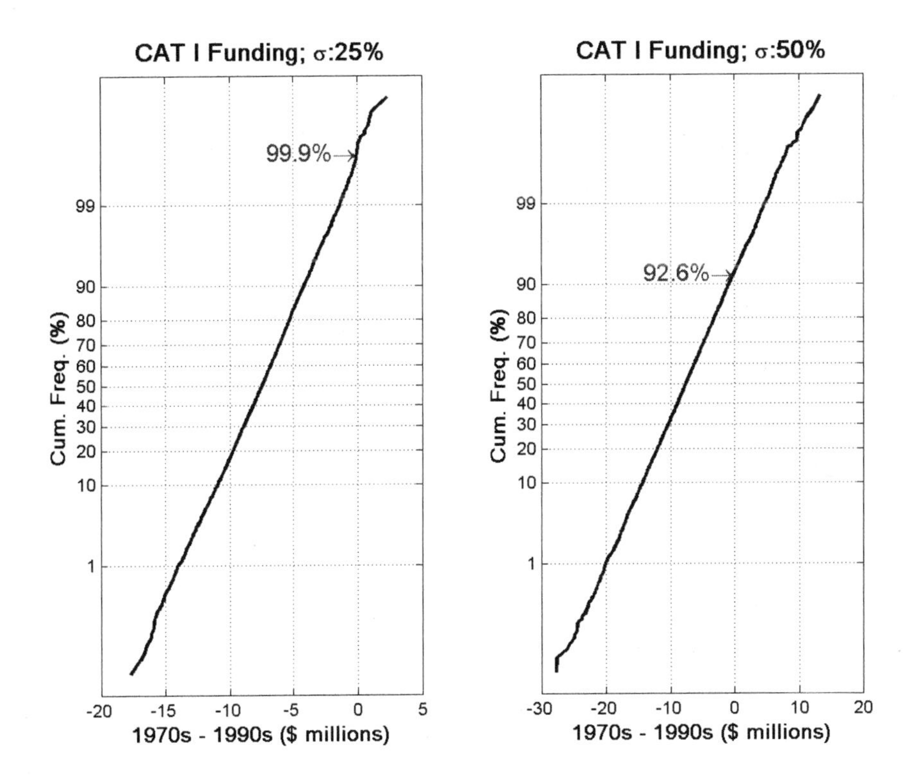

FIGURE C-2 Cumulative frequency distributions of 10,000 samples of generated differences between means of years 1973, 1974, 1975 and of years 1999, 2000, 2001 for Category I funding.

CATEGORY II (WATER CYCLE)

As shown in Figure C-3, the late 1990s level funding is significantly higher than the mid 1970s level funding, especially for a best-guess standard deviation level (25 percent of mean).

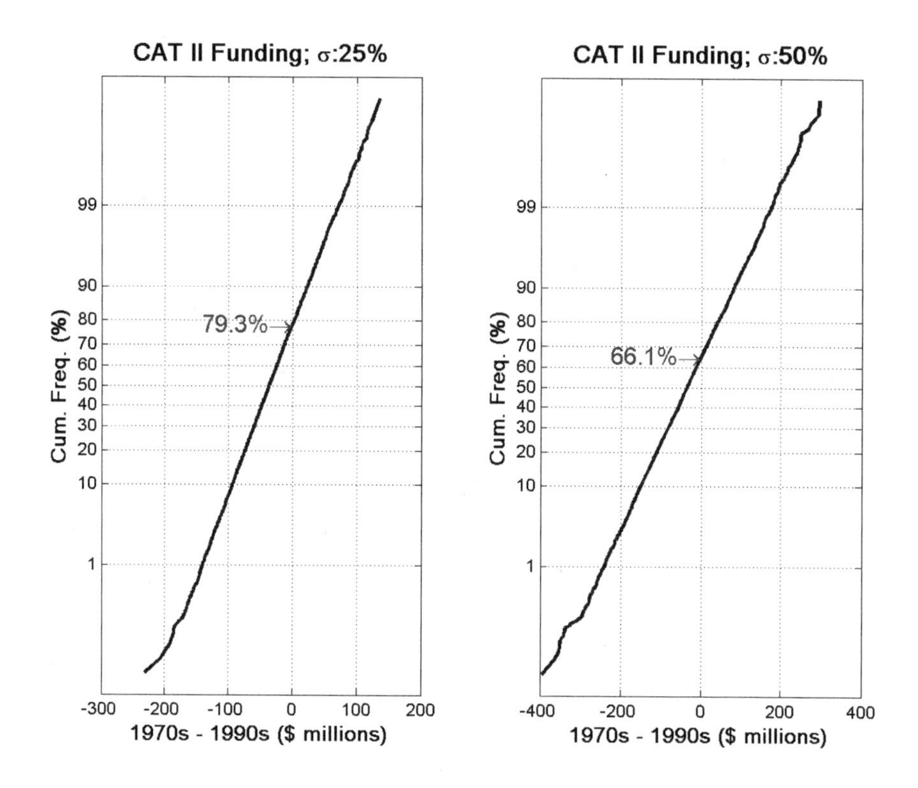

FIGURE C-3 Cumulative frequency distributions of 10,000 samples of generated differences between means of years 1973, 1974, 1975 and of years 1999, 2000, 2001 for Category II funding.

CATEGORY III (WATER SUPPLY AUGMENTATION AND CONSERVATION)

As shown in Figure C-4, the likelihood that the late 1990s level funding is higher than the mid 1970s level funding is very small in both cases of uncertainty depicted, and one concludes that funding has declined substantially in the late 1990s with respect to the mid 1970s for this category.

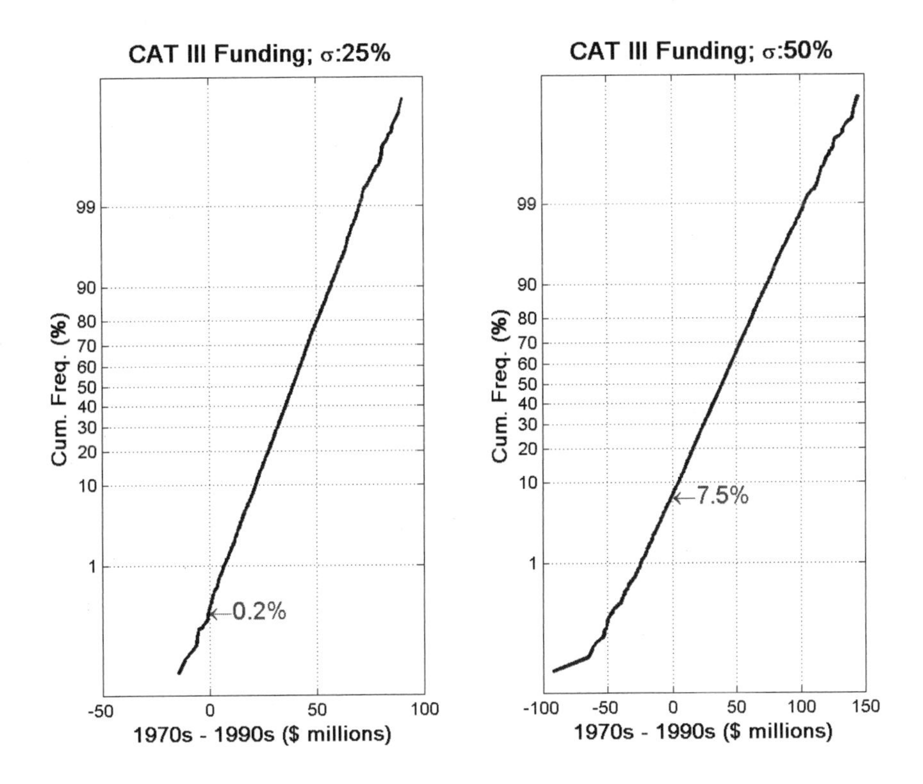

FIGURE C-4 Cumulative frequency distributions of 10,000 samples of generated differences between means of years 1973, 1974, 1975 and of years 1999, 2000, 2001 for Category III funding.

CATEGORY IV (WATER QUANTITY MANAGEMENT AND CONTROL)

As shown in Figure C-5, this is another case for which the mid 1970s funding level and the late 1990s funding level are not significantly different.

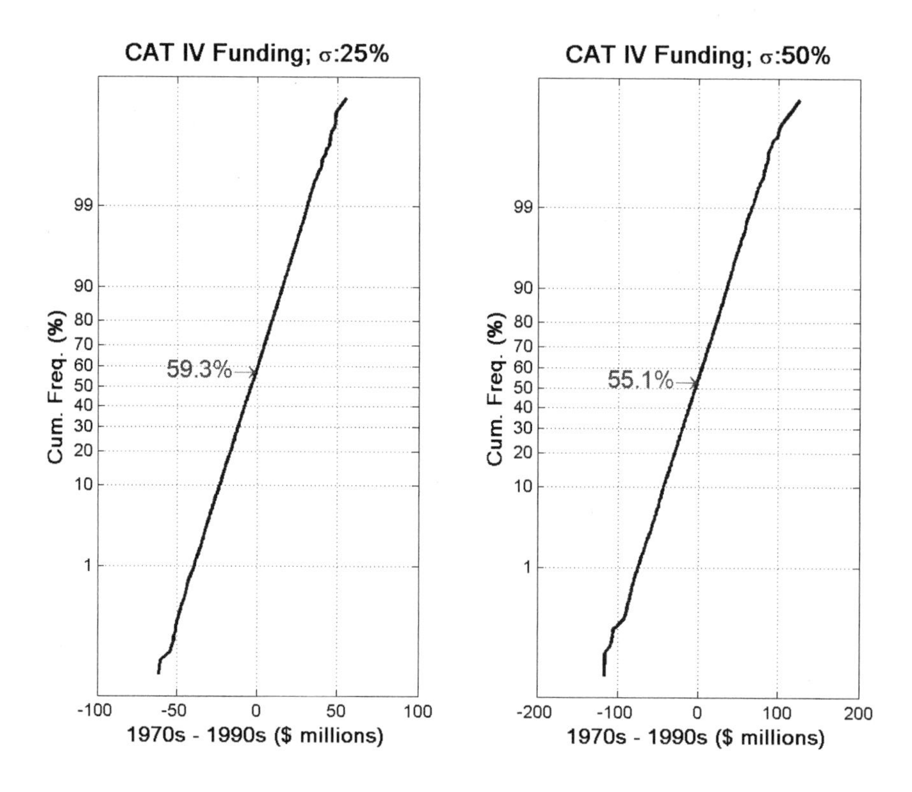

FIGURE C-5 Cumulative frequency distributions of 10,000 samples of generated differences between means of years 1973, 1974, 1975 and of years 1999, 2000, 2001 for Category IV funding.

CATEGORY V (WATER QUALITY MANAGEMENT AND PROTECTION)

As shown in Figure C-6, for this category, the mid 1970s funding is higher than the late 1990s funding with high confidence.

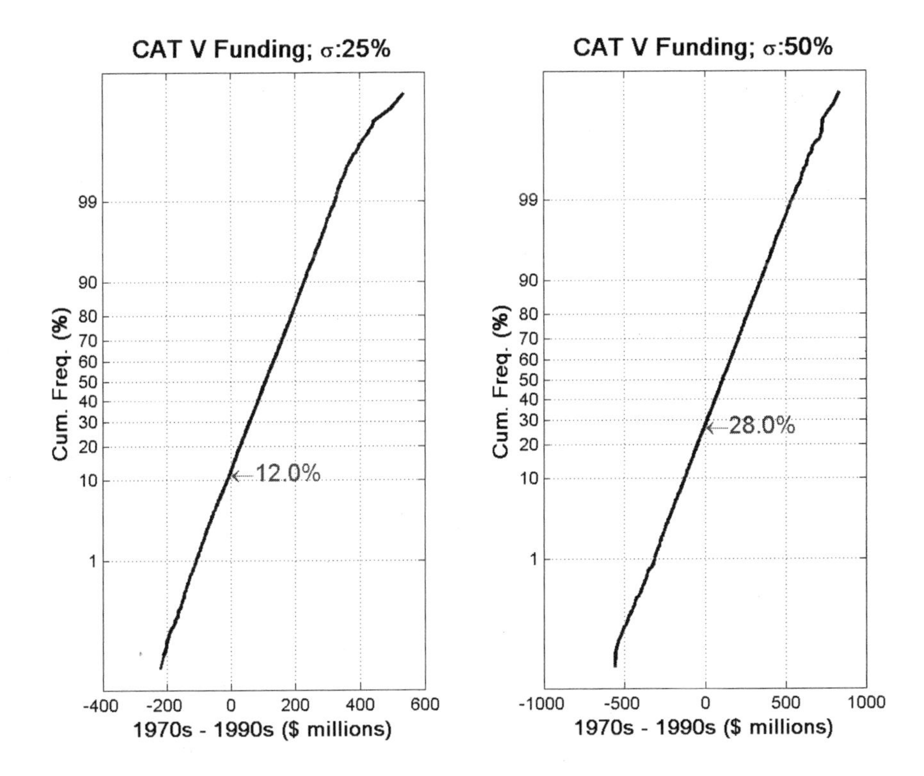

FIGURE C-6 Cumulative frequency distributions of 10,000 samples of generated differences between means of years 1973, 1974, 1975 and of years 1999, 2000, 2001 for Category V funding.

CATEGORY VI (WATER RESOURCES PLANNING AND OTHER INSTITUTIONAL ISSUES)

For this category, Figure C-7 shows that the mid 1970s funding is higher than the late 1990s funding with high confidence.

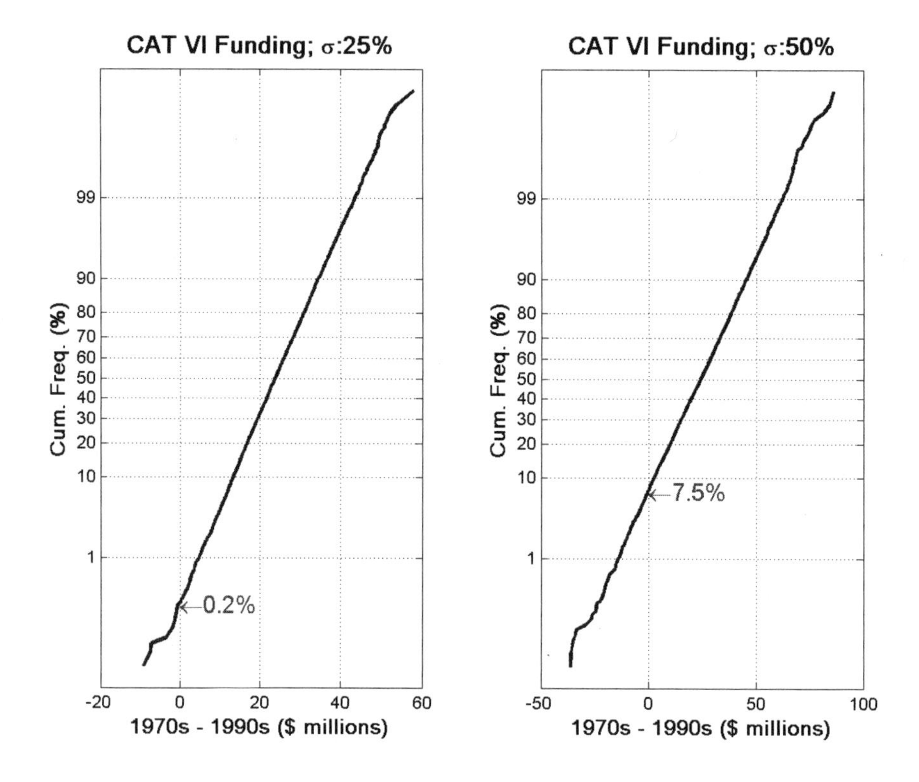

FIGURE C-7 Cumulative frequency distributions of 10,000 samples of generated differences between means of years 1973, 1974, 1975 and of years 1999, 2000, 2001 for Category VI funding.

CATEGORY VII (RESOURCES DATA)

As shown in Figure C-8, for Category VII the mid 1970s funding is higher than the late 1990s funding with high confidence.

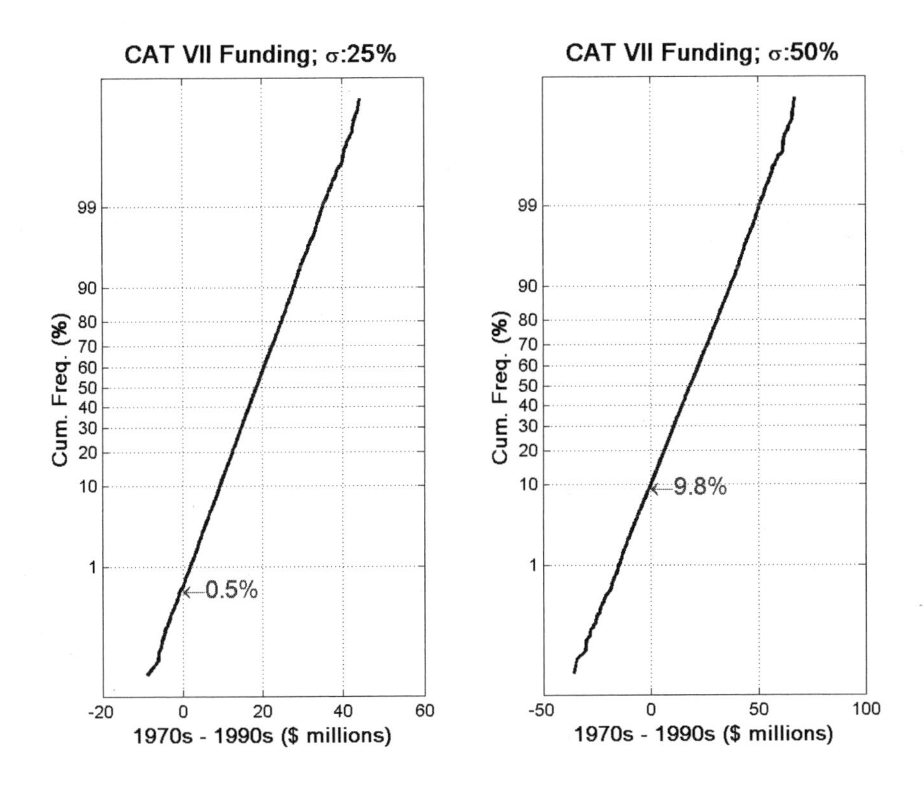

FIGURE C-8 Cumulative frequency distributions of 10,000 samples of generated differences between means of years 1973, 1974, 1975 and of years 1999, 2000, 2001 for Category VII funding.

CATEGORY VIII (ENGINEERING WORKS)

For this category, the assumed uncertainty makes a difference, as shown in Figure C-9. If it is assumed at the high value (standard deviation equal to 50 percent of the mean), the funding levels in the early (mid 1970s) and recent (late 1990s) periods cannot be distinguished with high confidence. However, if the best guess of uncertainty (25 percent is used to compute standard deviation) is assumed valid, the late 1990s funding is significantly higher than the mid 1970s funding.

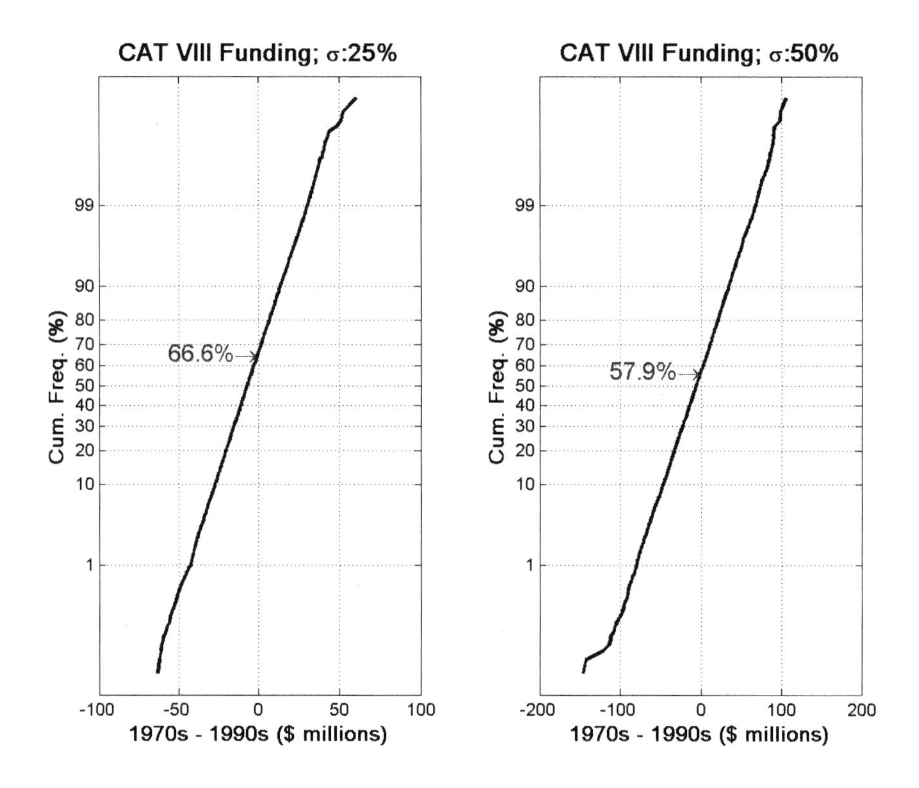

FIGURE C-9 Cumulative frequency distributions of 10,000 samples of generated differences between means of years 1973, 1974, 1975 and of years 1999, 2000, 2001 for Category VIII funding.

CATEGORY IX (MANPOWER, GRANTS, AND FACILITIES)

As shown in Figure C-10, in this case, the late 1990s funding level is substantially higher than the mid 1970s funding level with high confidence.

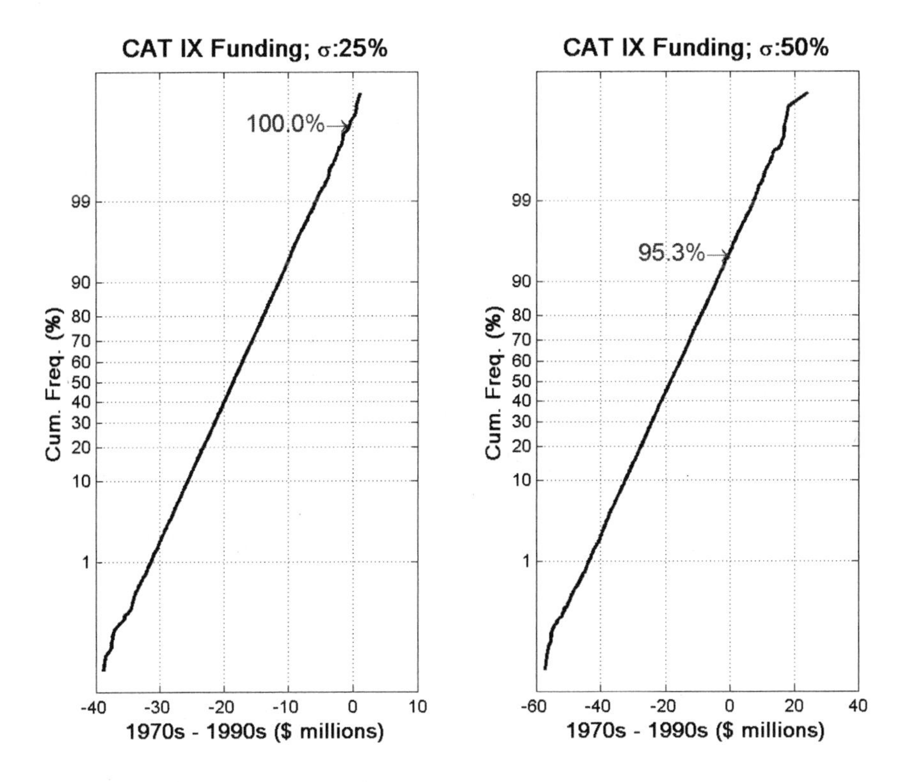

FIGURE C-10 Cumulative frequency distributions of 10,000 samples of generated differences between means of years 1973, 1974, 1975 and of years 1999, 2000, 2001 for Category IX funding.

CATEGORY X (SCIENTIFIC AND TECHNICAL INFORMATION)

As shown in Figure C-11, for this category, the mid 1970s funding is higher than the late 1990s funding with high confidence.

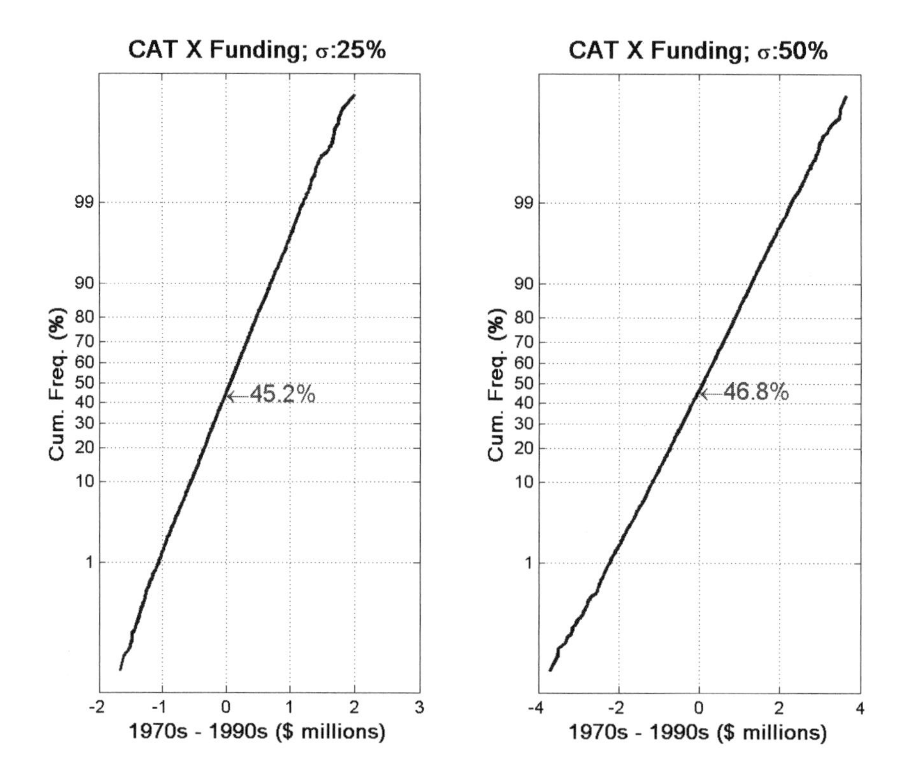

FIGURE C-11 Cumulative frequency distributions of 10,000 samples of generated differences between means of years 1973, 1974, 1975 and of years 1999, 2000, 2001 for Category X funding.

CATEGORY XI (AQUATIC ECOSYSTEM MANAGEMENT AND PROTECTION)

For this category, Figure C-12 shows that the late 1990s funding is much higher than the mid 1970s funding with high confidence (note that for the 25 percent uncertainty case, the zero-level cumulative probability distribution of the differences is 100 percent, as the entire curve lies on the negative x-axis half).

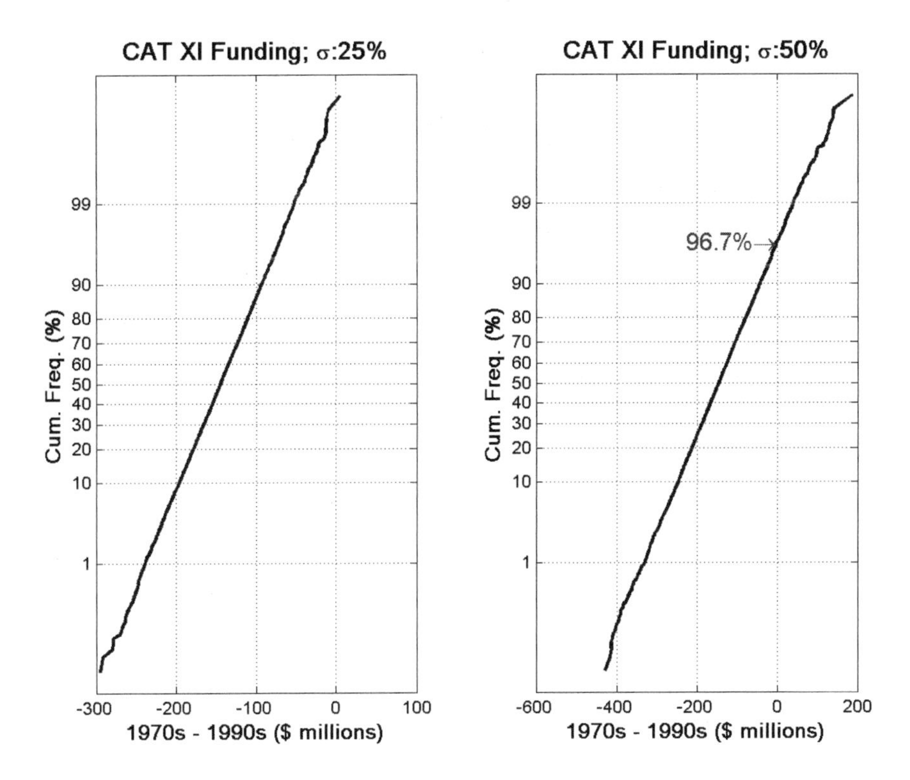

FIGURE C-12 Cumulative frequency distributions of 10,000 samples of generated differences between means of years 1973, 1974, 1975 and of years 1999, 2000, 2001 for Category XI funding.

TOTAL WATER RESOURCES RESEARCH FUNDING MINUS CATEGORY XI (AQUATIC ECOSYSTEM MANAGEMENT AND PROTECTION) FUNDING

When funding for Category XI is subtracted from the total water resources research funding, the late 1990s funding is lower than the mid 1970s funding with high confidence, even when the uncertainty in funding levels is 50 percent (see Figure C-13).

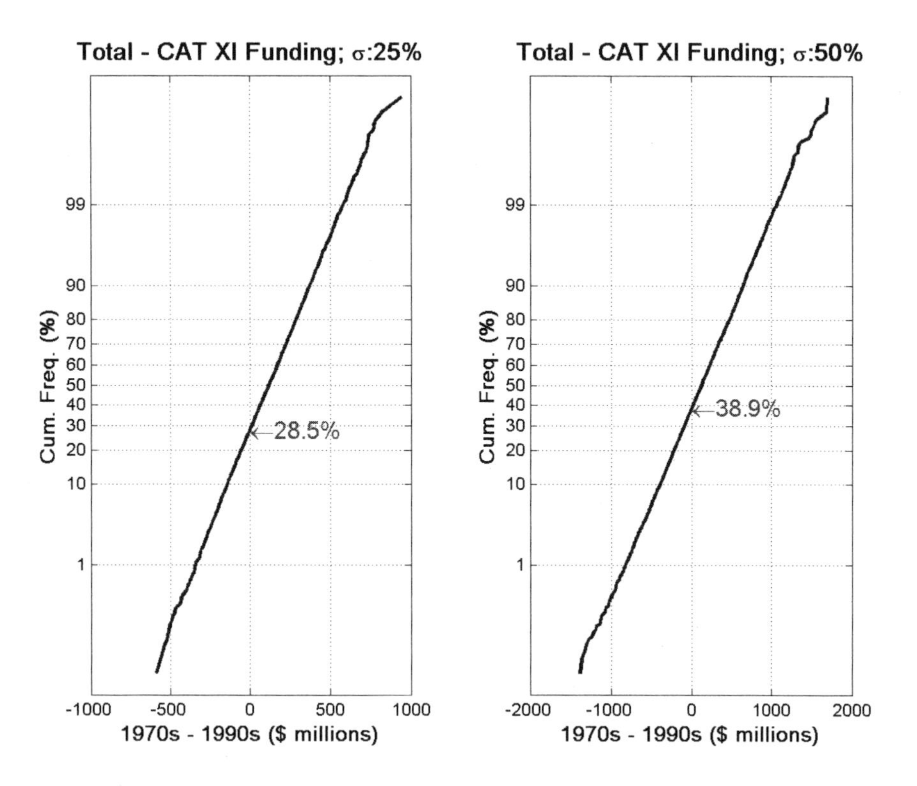

FIGURE C-13 Cumulative frequency distributions of 10,000 samples of generated differences between means of years 1973, 1974, 1975 and of years 1999, 2000, 2001 for the total water resources research funding minus Category XI funding.

Appendix D

Summary of State Perspectives

On January 9, 2003, the Committee on Assessment of Water Resources Research met in Tucson, Arizona, and heard eight presentations about individual state's water resources research needs and water issues. Subsequently, five additional states were contacted to fill out the geographical representation of the submissions. The goal of the exercise was twofold. First, the committee wanted the states to have an opportunity to express their needs with respect to water resources research, since they are often the primary user of the results of federally funded research activities. Second, the committee used the information to confirm that the research needs found in *Envisioning the Agenda for Water Resources Research in the Twenty-first Century* actually reflect current thinking at the state level. The 13 individuals and their state affiliation are listed below.

1. Rita McGuire, Arizona Center for Public Policy
2. Steve Macaulay, California Department of Water Resources
3. Karl Dreher, Idaho Department of Water Resources
4. Derek Winstanley, Illinois State Water Survey
5. Steve Kahl, Maine Water Resources Research Institute
6. Mark Smith, Massachusetts Department of Environmental Affairs
7. Mark Buehler, Metropolitan Water District of Southern California
8. Jamie Crawford, Mississippi Department of Environmental Quality
9. Roger Patterson, Nebraska Department of Natural Resources
10. Peggy Barroll, New Mexico Office of the State Engineer
11. Barry Norris, Oregon Water Resources Department
12. Andrew Zemba, Pennsylvania Office of Water Management
13. Bill Mullican, Texas Water Development Board

Each individual was asked to provide input on three water issues, as listed in boldface below. Instead of presenting the individual responses of each, what follows are brief summaries of all of the responses. For a transcript of a complete state response or a copy of the PowerPoint presentations, contact Laura Ehlers at lehlers@nas.edu.

1. *Provide a brief description of your organization's responsibilities.*

The 12 contributing organizations include state agencies responsible for water administration, public water supply, planning, and data gathering as well as agencies responsible for water and public policy research and an agency responsible for providing metropolitan water supplies.

2. *Speaking from the perspective of your state and its water management institutions, what are the most important issues that you are likely to face in the next 10–15 years? Please do not discuss short-term operational problems.*

In many instances it was difficult to distinguish a state's water issues from the water resources research that is needed to address them. Nonetheless, it was clear that important issues span a wide range of topics. Those issues/topics that were mentioned by more than one state include

- continuing need for better data collection
- meeting the goals of the Endangered Species Act
- dealing with future climate change
- how to manage groundwater mining
- how to take surface water–groundwater interactions into account when setting policy
- dealing with droughts and floods
- capturing recharge
- various water quality issues, particularly emerging contaminants and pathogens

Other issues of concern to the states include interstate compact compliance issues, adjudication of water rights, dam safety/aging structures and finding cost-effective ways to deal with infrastructure, vegetation management, land subsidence due to water withdrawals, sedimentation of reservoirs, growing water demand, treatment and disposal of brine from desalination plants, exotic species invasions, and Total Maximum Daily Loads and the general problem of nonpoint source pollution.

3. *What kinds of research would be most helpful in providing the knowledge and technology needed to address these long-term issues?*

Below are abridged responses listed by the category of research. No attribution is given for the contributing agency because in a number of instances the same or very similar suggestions were made by multiple agencies. However, in other instances, the comment reflects the views of only one agency representative. Note that almost all of the state representatives mentioned the need for better hydrologic data, even though it was explained that the committee and this report are not considering data collection per se to be a research activity. It is acknowledged in Chapter 5 that research efforts are often limited by the availability of high-quality data.

Data Collection

- Better and more reliable stream gaging is needed.
- Better basic hydrologic data are needed, especially in real time.
- Research is needed on how to improve stream gaging, e.g., by using remote sensing and tomographic methods.
- Soil moisture data collection, as well as research to improve this, is needed.
- Monitoring of land subsidence was voiced as important but currently neglected.
- The extent and location of impervious surfaces need to be determined.

Endangered Species

- Better science is needed to understand the water needs of listed endangered species, both aquatic and terrestrial. This comment was made by a large number of the participating state representatives, although each mentioned a different species.

Surface Water–Groundwater Interactions

- A better understanding of surface water–groundwater interactions is needed to help the states answer such questions as:

1. *What have been the consequences of ignoring the long-term effects of developing groundwater that is connected to fully appropriated surface water?*
2. *Is there an economic/ecological time horizon for consideration of interference with surface water, or is the establishment of hydraulic connection without regard to time and percent of interference sufficient to determine injury to senior rights?*
3. *What methods are available, or could be developed, to give rapid reasonable estimates of return flow and consumptive uses?*

It was suggested that a comprehensive survey of surface water–groundwater interactions along major aquifer systems would be useful.

Tools to Aid in Water Management

- Better hydrologic models are needed, for both surface water and groundwater.
- Research is needed on coupling climate models to models used to predict surface water availability.
- Improved models are needed for predicting surface water availability (runoff forecasting) at different time scales.
- Models that could provide an assessment of worst-case drought and flood scenarios would be useful.
- Research is needed on real-time water management decision making.
- Research is needed on how to accomplish water transfers.
- Research is needed on how to implement and make successful programs on water demand management. For example, what are the tools needed to promote the adoption of conservation practices?
- Planning is needed for aging dam structures.
- A better understanding is needed on how to mitigate sabotage of water systems.

Water Conservation/Recharge/Drought

- Research is needed on wastewater reuse, particularly the required treatment.
- Research is needed on recycling and effective use of gray water and stormwater.
- Research is needed to develop more low-water-use crops.
- Social science research is needed on the severity of drought impacts and what institutional responses should be.
- Research is needed to better understand and quantify recharge, to develop recharge monitoring techniques, and to understand the effects of human activity on groundwater recharge (e.g., benefits of stormwater best management practices).

Water Quality

- Research is needed on the water quality impacts of wastewater effluent.
- Models of the fate and transport of surface water and groundwater pollutants need to be developed, particularly for nonpoint sources of pollution.
- Research is needed on desalination and the environmental impacts of brine disposal.
- More information is needed on potential point-of-use devices.
- Research is needed on alternate methods of preventing sedimentation of reservoirs.
- Research is needed on the economic and environmental impacts of nutrient pollution and on alternative nondischarge uses of nutrients.

Vegetation

- Research is needed on vegetation management and how it affects overland flow and groundwater recharge.
- Research is needed on the use of remote sensing to address vegetative management and stream gaging, particularly in ephemeral streams.
- Information is needed on the long-term impacts of invasive species (e.g., aquatic weeds) and how to prevent, control, and eradicate unwanted vegetation.
- Research is needed on how landscape changes brought about by human activity affect water quantity and quality.

COMMITTEE OBSERVATIONS

The following observations were made by the committee following the presentations:

- States are no longer receiving federal money for collecting hydrologic data or receiving the data itself that can be applied to local problems. Thus, basic data collection has become an unfunded mandate for the states.
- State representatives feel that where the federal government owns a great deal of land (notably in the West), it needs to take the lead research role.
- Data collection at the federal level is also needed for consistency purposes.
- Meeting human water supply needs while meeting environmental needs will present challenges in almost all regions of the country.
- The Water Resources Research Institutes provide broad advantages and increase the stature of basic data but receive low federal funding.
- Many of the states' representatives expressed disaffection with the federal water resources research enterprise. Thus, there need to be better linkages between the federal programs that generate and fund research and the state users of such research.

Appendix E

Charter of the Subcommittee on Water Availability and Quality Committee on Environment and Natural Resources National Science and Technology Council

A. Official Designation

The Subcommittee on Water Availability and Quality (Subcommittee) is hereby established by action of the National Science and Technology Council (NSTC) Committee on Environment and Natural Resources (Committee).

B. Purpose and Scope

The purpose of the Subcommittee is to advise and assist the Committee and the NSTC on policies, procedures, plans, issues, scientific developments, and research needs related to the availability and quality of water resources of the United States. For the purpose of this Subcommittee, water resources are defined as fresh and brackish water in the atmosphere, streams, lakes, unsaturated zone, aquifers, and estuaries. The Subcommittee will focus on science issues and policy related to needed improvements in technology and research that will advance the goal of ensuring a safe and sustainable supply of water in the United States for human and ecological needs.

C. Functions

To advance its goal, the Subcommittee will carry out the following functions:

1. Facilitate communication and coordination among federal agencies and representatives from nonfederal sectors on issues of science, technology, and policy related to water availability and quality.

2. Advise the Committee on significant recent developments in science and technology related to the assessment and enhancement of water availability and quality.

3. Determine needs for additional research, monitoring, and technology development.

4. Develop a plan for a coordinated multiagency effort to provide the needed research, monitoring, and development.

5. Recommend budget priorities that will target federal spending toward the most critical needs for ensuring safe and sustainable water supplies for human and ecological uses.

6. Provide reviews and analyses of federal policies and programs that affect water availability and quality.

7. Advise the Committee on linkages between the availability and quality of water and the nation's economic and strategic security.

8. Assess periodically (a) priorities for research and development of systems related to enhancement of water supplies, and (b) research and development of systems related to monitoring and forecasting water flow and quality and their effect on aquatic life.

D. Membership

Members of the Subcommittee will be from the following agencies:

Council on Environmental Quality
Office of Management and Budget
Office of Science and Technology Policy
National Aeronautics and Space Administration
National Oceanic and Atmospheric Administration
National Park Service
National Science Foundation
Tennessee Valley Authority
U.S. Agricultural Research Service
U.S. Army Corps of Engineers
U.S. Bureau of Reclamation
U.S. Department of Energy
U.S. Department of Health and Human Services
U.S. Environmental Protection Agency
U.S. Fish and Wildlife Service
U.S. Forest Service
U.S. Geological Survey
U.S. Natural Resources Conservation Service

E. Private Sector Interface

The Subcommittee may seek advice from the President's Committee of Advisors on Science and Technology (PCAST) and will recommend to the Assistant to the President for Science and Technology the nature of additional private-sector advice needed to accomplish its mission. The Subcommittee may also interact with and receive ad hoc advice from various nonfederal groups as consistent with the Federal Advisory Committee Act, such as the Advisory Committee on Water Information.

F. Duration

The Subcommittee shall serve for a duration of 5 years, until September 30, 2007. The charter, however, may be renewed by the Committee and the chairman of the NSTC at their discretion.

G. Approval

By my signature below, I hereby approve the formation of the Subcommittee on Water Availability and Quality, subject to the terms in this charter, as a function of the Executive Branch consistent with the public interest and with its lawful duties.

Director, Office of Science and Technology Policy Date

THE CHALLENGE OF THE "THIRD ERA OF WATER RESOURCES"

A paper presented to the Subcommittee on Water Availability and Quality (SWAQ) by its co-chairs, May 5, 2003
(Revised June 10, 2003)

Science and technology has always been crucial to the proper development and protection of our nation's water resources. We would propose that since about the middle of the 1990's the nation has entered into a "third era of water resources." And it is the realities of that third era that must form the basis of our science and technology agenda. So, what are these three eras?

The first era spans the time from the first development of cities, industry, and irrigated agriculture up to about 1972. During this era the resource was developed through the building of dams for purposes of water supply, navigation, hydroelectric power generation and flood control. The key science question behind these projects was about the amount of water, or power, that could reliably be delivered from these projects. Groundwater development only became a significant factor in water development after the invention of center pivot irrigation systems and high-capacity submersible pumps in the middle of the 20th century. In only a few cases did groundwater development lead to long-lasting or far-reaching impacts during this era. The science question in groundwater development was about the amount of water that could be extracted from a set of wells without unduly affecting neighboring users. Waste from industry and cities were generally discharged to rivers with little or no treatment. At the end of this era public concern over pollution was growing rapidly and causing scientists to address difficult questions about how multiple sources of pollution along a river were each affecting water quality or the biota. The goal of this science was to assign responsibility appropriately and to make decisions about the most effective means of improving water quality. This latter task was difficult given the state of scientific understanding, monitoring technology, and computational capability at the time, and results were often contended in legal proceedings, leading to major logjams in solving the nation's water pollution problems.

The second era starts in 1972 and runs to the middle 1990s. During this era there were very few new water storage projects built. Any expansions in deliveries of surface water came about by simply extracting more from the infrastructure already built. In those areas where water use was growing, the West and the South primarily, this resulted in growing stresses on the total supply for the users and in significant impacts on the aquatic community, as these off-stream diversions left little water in the rivers during dry periods. Groundwater use increased rapidly due to better technology and because of the limits on surface-water supplies. However, during the same period, contamination of groundwater became a concern and a focus for remediation. The Clean Water Act resulted in major enhancements in treatment of municipal and industrial waste. It was a period of very active development of treatment technology, but little attention was paid to the science of water quality, because the Clean Water Act placed its focus for the first two decades on application of technology to clean up point sources.

The third era began in the mid 1990s and can be expected to run for many more years. Many regions of the nation are now facing clear and intense conflict among the major categories of use: urban, industrial, agricultural, and ecological. Because large-scale water transfers or building of additional reservoirs are unlikely, the interests in most regions are actively involved in some kind of renegotiation over the uses of the existing supplies. The key science question is now not how much water can we reliably deliver from the river, but rather how much water do we need to leave in the river for ecosystem functions. There is interest in new systems for storing water, but these are primarily systems for capturing surface water in times of high flow and storing it in aquifers to be extracted months or years in the future during times of need. These systems represent an important emerging technology, but one whose performance is not well known at this time. Groundwater development is increasing rapidly because of the limitations on further surface-water sources. What is becoming more apparent is that withdrawal of groundwater can have significant impacts on surface water and on aquatic ecosystems over time scales of decades and spatial scales of tens of miles. Understanding groundwater and its connection to surface water thus becomes a crucial science need for wise management of the resource. Finally, in the area of water quality, the improvements due to the technology-based approach have been significant, and yet problems remain for the water quality and the aquatic biota. These problems are significantly related to land uses, particularly agricultural and urban land management practices. Future decisions about land use and land management practices must be based on a predictive understanding, supported by empirical data, of the relationship between those activities and the water quality and biological end points. These impacts must be understood at the scale of the local watershed, but also at scales of major river basins with impacts persisting hundreds of miles or more downstream from their sources.

SCIENCE AND TECHNOLOGY FOR THE THIRD ERA OF WATER RESOURCES

The era of plentiful, clean water supplies and pristine, biologically diverse aquatic resources has been replaced by an era of highly developed water resource management and regulatory and voluntary programs to restore and maintain water quality. Science and technology generated via federal research have enabled both the utilization of water resources and the information and technologies needed to guide related public policy and economic development needs. The United States, and the world at large, are now in a "third era of water resources" characterized by:

- increasing competition and conflict among users of increasingly scarce water resources, often leading to technical and policy gridlock because of the conflict over values, protracted litigation or regulatory actions, and the scientific uncertainties
- federal budget realities coupled to policy, and program performance expectations requiring scientifically based information and technology to inform decision making at all scales of government from federal to local. This information must fully explain the costs, effectiveness, and benefits of actions to provide water, prevent water quality impairment, and to restore water quality in impaired systems
- a shift from relying exclusively on "command and control" approaches to solve problems to approaches that harness market forces, provide flexibility and efficiency, and that build on principles of sustainable water quantity, water quality, and economic development
- a recognized shift from industrial and wastewater treatment discharges as the major cause of water quality impairment to nonpoint source runoff, atmospheric deposition, and groundwater inflows as the major pathways for contamination of water resources
- a scarcity of freshwater and increasing costs to store, extract, purify, and distribute water suitable for irrigation, drinking, basic sanitation, and aquatic habitat maintenance

URGENT AND NATIONAL PRIORITIES FOR THE THIRD ERA

The federal water science and technology programs exist to meet the nation's water availability and quality needs. These programs will be challenged by the emerging Third Era. The SWAQ will structure its work to see to it that the programs do not simply continue to pursue the issues of the past. Rather, we will work together to explore how the agencies can effectively make the significant progress, during the next five years, on some of the most urgent problems posed by the Third Era. The SWAQ has selected two issues for initial consideration and special study. Both are compelling, interagency, national, and policy-relevant priorities. Both will require a realignment of current priorities and may require new resources. The work of the SWAQ will not be limited to these two issues, but they are suggested by the co-chairs as initial topics for consideration.

1. Quantifying the future availability of freshwater in light of both withdrawal uses and ecosystem uses:

A very common problem in the Third Era is that the existing infrastructure for storing and delivering water for uses such as agricultural, urban, or industrial needs is now being called upon to support healthy biotic communities in rivers and associated lakes, wetlands, and floodplains. Also, groundwater development is causing declines in groundwater storage and this is resulting decreased baseflow in streams and increased water temperatures during critical times in the summer. Answering questions of future availability of freshwater requires an ability to predict at least four major types of variables. (1) There is a need to estimate the future economic demands for water for withdrawal uses (agriculture, urban, and industrial). The complexities of this process involve forecasting of changes in technology, economic activity, and the response of the legal and political system to shifting water from one type of use to another. (2) There is a need to estimate the response of the biological community to changes in streamflow and stream temperature, clarity, and chemistry. This question is often pivotal to addressing instream flow needs. (3) There is a need to estimate the degree to which aquifer storage is changing and will change in the future (given various land and water use patterns) and then how these changes in groundwater will affect the flow, temperature, and chemistry of streamflow. (4) There is finally a need to estimate how surface-water flow is changing as a result of water management activities, land-use change, climate change, diversions, and storage.

Scientific tools and data that are needed to respond to the questions of water availability include new approaches to estimating of groundwater recharge, discharge, and storage at a regional scale. This will demand the use of improved sensors such as ground-based and space-based gravity methods as well as tracer methods (using isotopes, heat, or man-made chemicals as tracers) and improved groundwater and groundwater–surface-water models. Efforts are needed to explore new water-using technologies and the economic and social barriers to their adoption by citizens, communities, industry, and farmers.

Traditional science and engineering of river management has focused on the question: "How much water can we reliably extract from this river or aquifer, given the systems of reservoirs, diversions and wells?" The new question is "How much water do we need to leave in the river or aquifer to support the biota?" Without scientifically defensible answers to this kind of question, regional water management decisions will remain gridlocked in a manner that serves neither the withdrawal users nor the ecosystem. The recent article in *Science* magazine regarding the Klamath River Basin gives a prime example of this gridlock, but many other examples exist nationwide (e.g. the California Bay Delta, Everglades, Grand Canyon, Platte River, Appalachacola-Chattahoochie-Flint, to name a few). The science that is needed is improved understanding of the stressor–response relationship between the physics and chemistry of managed river systems and the response of the biota. In order to move forward with resolution of these conflicts there needs to be research leading to improved models that can be used to diagnose the cause of current problems and predict the future state of the biota. The science needs to consider the wide range of factors that can affect the biota: including: streamflow, water temperature, sediment concentrations, water chemistry, riparian vegetation, groundwater development impacts on surface water flow and temperature, geomorphology, physical barriers, harvest, invasive species, disease, and dissolved gases.

In short, improving the scientific basis for water availability planning can only be done through an integrative effort involving surface-water and groundwater hydrology, climatology, water chemistry, water engineering and economics, and the biological responses of ecosystems to ongoing environmental change.

2. Assessing and predicting the effectiveness of land use practices and watershed restoration on water quality and ecosystem health.

Even though the nation has made great strides in reducing urban and industrial sources of pollution, the quality of many water bodies still does not support the level desired from the standpoints of protecting human health and healthy biotic communities. For example we know that

1. Many river, lake, estuarine, and near-coastal systems (e.g., Gulf hypoxia) are impaired by nutrient enrichment and that the primary sources for these nutrients are from agricultural and urban uses and from atmospheric nutrient inputs.
2. Pathogenic organisms are common in our nation's waters, and they threaten recreational water use and safe drinking water supplies. Sources include sewer overflows, leaky sanitary sewers, malfunctioning septic tanks, animal production facilities, pets, and wildlife.
3. Sediment originating from erosion of the landscape continues to create problems in aquatic systems. It causes direct harm to fish and shellfish, and often carries toxic metals or organic chemicals.
4. Mercury is a significant problem to the higher-level organisms in the aquatic food chain (fish and fish-eating birds and mammals, including humans who subsist on fish). The source of this mercury is often atmospheric, and the likely effectiveness of control strategies is poorly known at this time.

A common characteristic of most of these water quality problems is that they require significant interventions in land use practices at watershed or large river-basin scale. These interventions, commonly called "best management practices" (BMPs), ecosystem restoration practices, and watershed management action programs are expected to yield their results very slowly. Because these practices involve changing the movement of water, chemicals, and sediment through soil and through groundwater and over long distances within a watershed, the desired outcomes may take decades to emerge. Documenting the effectiveness of these measures on water quality or biological end points is confounded by the significant temporal variation that comes from the natural variation between seasons and between wet years and dry years. Science and technology must find the "signal" of effectiveness amidst the considerable natural "noise" of the watershed system. The challenge to the water science and technology community is to demonstrate and ultimately predict the costs, effectiveness, and the benefits of the various strategies that are being deployed to restore or improve the nation's aquatic systems.

HOW SWAQ WILL ADDRESS THESE URGENT AND NATIONAL PRIORITIES FOR THE THIRD ERA

We propose that the SWAQ undertake a review that describes the importance of these two problem to the nation's economic development and ecosystem conditions, the status of knowledge, current research efforts and then describe the kind of research and development needed to move the science forward to resolve these kinds of problems. Special attention will be given to data needs, integrated modeling needs, and the needs for new sensors that can contribute to this overall effort.

Getting the metrics right: measuring, modeling and the adaptive management feedback loop

Actions taken to improve water quality or ecosystem health must be viewed, in part, as experiments. We are engaged in a process of adaptive management where we take actions that should move us in the right direction, but also recognize that there is great uncertainty about how well those actions will succeed in producing the desired outcomes. Any good experimental design demands a rigorous measurement plan, data analysis plan, hypothesis tests, and reporting of results in the peer reviewed literature. The SWAQ will explore how well the nation is positioned to conduct adaptive management of water, with the complete feedback loop from design, to action, to data, to results, and back to design and action. The SWAQ will pose the following kinds of questions as it pursues the two issues proposed above for initial study:

1. What are the metrics required to assess the costs, effectiveness, and benefits of approaches to meet the goals of the new strategies for water availability and quality?
2. Do existing sampling networks, statistical designs, and data collection programs use appropriate metrics, or can they be modified or expanded upon to measure across this integrated question of costs, effectiveness, and benefits?
3. What new technologies are needed to observe and measure the important variables in a cost-effective and timely manner, to serve the needs of adaptive management at local, regional, and national scales?

4. What is the most cost-effective way to use the current network of experimental watershed and groundwater study sites and related research programs to address the integrated questions of the Third Era?

5. What alternative experimental designs, survey approaches, and network-based approaches are needed to answer integrated questions and what institutional arrangements (including a role for the private sector) are required to implement targeted but large-scale "experiments"?

6. Is it possible to link existing models in order to link controls (such as BMPs) to ambient water quality to biological water quality (ecological end points) to economic benefits (such as monetized ecosystem goods and services)?

7. Are linked models credible? What is the best way to include social science algorithms?

8. What new models are required and what data collection is needed to test and implement the models?

ON-TIME DELIVERY OF VERIFIED TECHNOLOGIES

The Third Era is different from the first two water resources eras in that a robust private sector that develops, markets, and deploys technologies to meet water resources needs exists, and the capability and capacity are apparently available to further respond. The federal research community's role needs to shift from a development role to one of catalyzing the marketplace by identifying unmet needs. Another vital role is the issue of demonstration of performance and third-party verification. Put simply, as vendors offer leading edge technologies for remote sensing, monitoring, measurement, water treatment and desalination, and nanotechnology-based hardware, then users and buyers need data on the performance of the approaches. There are a number of models for this approach to draw from, and range from Cooperative Research and Development Agreements (CRADAs) to technology demonstration and verification programs. Specific issues that must be addressed are the following:

- Identification of a "hot needs" list for technologies needed to enable solutions to the SWAQ agenda
- An alignment of the focus of current technology research, development, demonstration, and verification programs with the SWAQ special emphasis areas

Action Items: The co-chairs propose the following:

1. Propose the concepts in this paper to the SWAQ and seek consensus on the specific initial topics both in number and scope.
2. Develop an action plan for the SWAQ that reflects the respective priorities and roles of the member agencies.

3. Characterize the content and expected results of the action plan and formally transmit reports with recommendations to the Committee on Environment and Natural Resources (CENR).

4. Complete first drafts of these two reports by the end of September.

Appendix F

Federal Agency and Nongovernmental Organization Liaisons

U.S. Department of Agriculture
Dale Bucks, Agricultural Research Service
Lisa Duriancik, Cooperative State Research, Education, and Extension Service
Carol Jones, Economic Research Service
Sheryl Kunickis, Natural Resources Conservation Service
Mike O'Neill, Cooperative State Research, Education, and Extension Service
Doug Ryan, U.S. Forest Service

U.S. Department of Commerce—National Oceanic and Atmospheric Administration
Gary Carter, National Weather Service
Richard Lawford, Office of Oceanic and Atmospheric Research
Donald Scavia, National Ocean Service

U.S. Department of Defense
Linda Chrisey, Office of Naval Research
Andrea Leeson, Strategic Environmental Research and Development Program
David Mathis, U.S. Army Corps of Engineers

U.S. Department of Energy
Teresa Fryberger, DOE headquarters
Henry Shaw, Lawrence Livermore National Laboratory

U.S. Department of the Interior
Mary Jo Baedecker, U.S. Geological Survey
Shannon Cunniff, U.S. Bureau of Reclamation
Chuck Hennig, U.S. Bureau of Reclamation

U.S. Department of Health and Human Services
Lorrie Backer, Centers for Disease Control and Prevention
Aaron Blair, National Cancer Institute
John Bucher, National Institute of Environmental Health Sciences
Robert Spengler, Agency for Toxic Substances and Disease Registry

U.S. Environmental Protection Agency
Anthony Maciorowski, Office of Water
John Reyna, Office of Research and Development
Molly Whitworth, Office of Research and Development

National Aeronautics and Space Administration
Jared Entin

National Science Foundation
Nick Clesceri, Engineering Directorate
Douglas James, Geosciences Directorate

General Accounting Office
Steve Elstein
Brenda Patterson

Office of Management and Budget
Jason Freihage

Water Environment Research Foundation
Jamie Montgomery

American Water Works Association Research Foundation
Chris Rayburn

The Nature Conservancy
Brian Richter

Water Resources Research Institutes
John Warwick, Nevada
Dave DeWalle, Pennsylvania
Ellen Weichert, Texas
Jan Urroz, Utah

Appendix G

Acronyms

AEC	Atomic Energy Commission
ARS	Agricultural Research Service (USDA)
AAAS	American Association for the Advancement of Science
AHPS	Advanced Hydrologic Prediction Service
ASLO	American Society of Limnology and Oceanography
ATSDR	Agency for Toxic Substances and Disease Registry (DHHS)
AWWARF	American Water Works Association Research Foundation
BLM	U.S. Bureau of Land Management (DOI)
BMPs	best management practices
CALFED	California Bay Delta Authority
CDC	Centers for Disease Control and Prevention (DHHS)
CENR	Committee on Environment and Natural Resources
CERCLA	Comprehensive Environmental Response, Compensation, and Liability Act
Corps	U.S. Army Corps of Engineers
COWRR	Committee on Water Resources Research
CSREES	Cooperative State Research, Education, and Extension Service (USDA)
CRADA	Cooperative Research and Development Agreement
CWA	Clean Water Act
DHHS	U.S. Department of Health and Human Services
DoD	U.S. Department of Defense

DOE	U.S. Department of Energy
DOI	U.S. Department of Interior
EDRP	Exposure-Dose Reconstruction Program
EMAP	Environmental Monitoring and Assessment Program
ENSO	El Niño Southern Oscillation
EPA	U.S. Environmental Protection Agency
ERS	Economic Research Service (USDA)
ESTCP	Environmental Security Technology Certification Program (DoD)
FCCSET	Federal Coordinating Council for Science, Engineering, and Technology
FCST	Federal Council for Science and Technology
FGDC	Federal Geographic Data Committee
FHA	Federal Housing Administration
FWS	U.S. Fish and Wildlife Service (DOI)
GAO	U.S. General Accounting Office
GDP	gross domestic product
GIS	Geographic Information Systems
GLERL	Great Lakes Environmental Research Laboratory (NOAA)
GLHHERP	Great Lakes Human Health Effects Research Program
GRACE	Gravity Recovery and Climate Experiment
HBN	Hydrologic Benchmark Network
HUD	U.S. Department of Health and Urban Development
HHS	U.S. Department of Health and Human Services
LTER	Long-Term Ecological Research
MTHM	Metric tons heavy metal
NASA	National Aeronautics and Space Administration
NAST	National Assessment Synthesis Team
NASQAN	National Stream Quality Accounting Network
NAWQA	National Assessment Water Quality program
NCEA	National Center for Environmental Assessment
NCER	National Center for Environmental Research
NCI	National Cancer Institute
NEHRP	National Earthquake Hazard Reduction Program
NERL	National Exposure Research Laboratory
NEXRAD	Next Generation Weather Radar
NHEERL	National Health and Environmental Effects Research Laboratory

NIEHS	National Institutes of Environmental Health Science
NIH	National Institutes of Health
NOAA	National Oceanic and Atmospheric Administration
NOS	National Ocean Service (NOAA)
NRC	National Research Council
NRMRL	National Risk Management Research Laboratory
NSB	National Science Board
NSF	National Science Foundation
NSDI	National Spatial Data Infrastructure
NSIP	National Streamflow Information Program
NSTC	National Science and Technology Council
NWS	National Weather Service (NOAA)
NWTRB	U.S. Nuclear Waste Technical Review Board
NWUIP	National Water-Use Information Program
OGP	Office of Global Programs (NOAA)
OMB	U.S. Office of Management and Budget
ONR	Office of Naval Research (DoD)
OOAR	Office of Oceanic and Atmospheric Research (NOAA)
ORD	Office of Research and Development (EPA)
OSTP	Office of Science and Technology Policy
OWRR	Office of Water Resources Research (formerly DOI)
PCAST	President's Council of Advisors on Science and Technology
PHS	U.S. Public Health Service
RCRA	Resource Conservation and Recovery Act
R&D	Research and development
SERDP	Strategic Environmental Research and Development Program
SDWA	Safe Drinking Water Act
STAR	Science Targeted to Achieve Results (EPA)
SWAQ	Subcommittee on Water Availability and Quality
TMDL	Total Maximum Daily Load
TNC	The Nature Conservancy
TSPA	Total System Performance Assessment
TVA	Tennessee Valley Authority
USBR	U.S. Bureau of Reclamation (DOI)
USDA	U.S. Department of Agriculture
USFS	U.S. Forest Service (USDA)
USGS	U.S. Geological Survey (DOI)

USGCRP	U.S. Global Change Research Program
WERF	Water Environment Research Foundation
WRB	Water Research Board
WRRIs	Water Resources Research Institutes
WSTB	Water Science and Technology Board

Appendix H

Biographical Sketches of Committee Members and Staff

Henry J. Vaux, Jr., *chair*, is a professor of resource economics, emeritus, at the University of California and associate vice president emeritus of the University of California system. He is currently affiliated with the Department of Agricultural and Resource Economics at the University of California, Berkeley. In addition to serving as associate vice president, Agriculture and Natural Resources, he previously served as director of the University of California Water Resource Center. His principal research interests are the economics of water use and water quality. Prior to joining the University of California, he worked at the Office of Management and Budget and served on the staff of the National Water Commission. He received a Ph.D. in economics from the University of Michigan. Dr. Vaux served on the NRC Committee on Western Water Management and the Committee on Ground Water Recharge, and he was chair of the Water Science and Technology Board from 1994 to 2001.

J. David Allan is a professor in the School of Natural Resources and Environment at the University of Michigan. He received a B.S. from the University of British Columbia and a Ph.D. from the University of Michigan. Dr. Allan is a member of the Ecological Society of America, the North American Benthological Society, and the American Society of Limnology and Oceanography. He has served on the board of editors of *Freshwater Biology* and the *Journal of the North American Benthological Society*. He serves as an advisor to American Rivers and The Nature Conservancy. His current research examines the influence of land use and landscape setting on the ecological status of streams and rivers, flow variability and its influence on the biological community, and indicators of stream ecosystem condition.

James Crook is an independent environmental engineering consultant specializing in the area of water reuse. He has previous experience in state government and consulting engineering arenas, where he developed and executed a broad range of engineering services for water and wastewater agencies in the public and private sectors in the United States and abroad. Dr. Crook developed California's first comprehensive water reuse criteria, has authored numerous technical papers and reports, and is an internationally recognized expert in the area of water reclamation and reuse. He was the principal author of water reuse guidelines published by the U.S. Environmental Protection Agency and U.S. Agency for International Development and was the American Academy of Environmental Engineers' 2002 Kappe Lecturer. He has served on the Water Science and Technology Board and several National Research Council committees. Dr. Crook received his B.S. in civil engineering from the University of Massachusetts and his M.S. and Ph.D. in environmental engineering from the University of Cincinnati.

Joan G. Ehrenfeld is a professor in the Department of Ecology, Evolution, and Natural Resources at Rutgers University. She is also the director of the New Jersey Water Resources Research Institute. She received her B.A. from Barnard College, Columbia University, magna cum laude with honors in biology, her M.A. in biology from Harvard University, and her Ph.D. in biology from City University of New York. Dr. Ehrenfeld is associate editor for the journal *Restoration Ecology* and has served on the editorial board of *Wetland*s. Her research is centered on the overlap between ecosystems ecology and plant ecology, emphasizing wetland ecology and exotic species invasions. She is involved in research in a wide variety of ecosystems in New Jersey, including the Pinelands, the hardwood forests of the northwestern hills, and the red maple swamps of the northeastern Piedmont province. Her teaching includes lecture courses on general ecology, wetland ecology, ecosystems ecology and global change, research methods in ecology, and restoration ecology.

Konstantine P. Georgakakos is the managing director of the Hydrologic Research Center in San Diego, California. He is also an adjunct full professor with the Scripps Institution of Oceanography of the University of California, San Diego, and an adjunct full professor with the Department of Civil and Environmental Engineering of the University of Iowa. Previously, he was an associate professor at the University of Iowa and with the Iowa Institute of Hydraulic Research, as well as a research hydrologist with the National Weather Service. He holds M.S. and Sc.D. degrees from the Massachusetts Institute of Technology. Honors and awards include the Presidential Young Investigator Award from the National Science Foundation and the NRC-NOAA Associateship Award from the National Research Council. He is the primary author of several software packages pertaining to real-time flow prediction, which are in various stages of implementation for operational use by the U.S. Army Corps of Engineers and the National

Weather Service. He is a consultant for the United Nations Food and Agriculture Organization, and he served as associate editor of the *ASCE Journal of Engineering Hydrology* and the *Journal of Hydrology*. He was elected the U.S. Expert in Hydrologic Modeling for the World Meteorological Organization Commission for Hydrology (1997–2000).

George R. Hallberg is a principal with the Cadmus Group, Inc., in Watertown, Massachusetts, conducting environmental research, regulatory analysis, and management services in the public sector. Previously he was associate director and chief of environmental research at the University of Iowa's environmental and public health laboratory and at the Iowa Department of Natural Resources. Dr. Hallberg is a long-time participant in NRC activities, including chairing the WSTB Committee on Opportunities to Improve the USGS National Water Quality Assessment Program. He has served on EPA's National Advisory Council for Environmental Policy and Technology and on the Office of Water's Management Advisory Group. Awards include the EPA Administrator's Award for Excellence in Pollution Prevention, the Soil Conservation Award from the Iowa Department of Agriculture and Land Stewardship, and the Distinguished Service Award from the Geological Society of America. His research interests have included environmental monitoring and assessment, agricultural-environmental impacts, fate and transport, contaminant occurrence and trends in drinking water, and health effects of environmental contaminants. Dr. Hallberg received a B.A. in geology from Augustana College and a Ph.D. in geology from the University of Iowa.

Debra S. Knopman is a senior engineer at RAND and associate director of RAND Science and Technology. Her expertise is in hydrology, environmental and natural resources policy, systems analysis, and public administration. From 1997 to 2003, she was a member of the Nuclear Waste Technical Review Board, which has oversight of the Yucca Mountain scientific and engineering program, and she chaired the Board's Site Characterization Panel. From 1995 to 2000, she served as director of the Center for Innovation and the Environment at the Progressive Policy Institute in Washington, D.C. She served as deputy assistant secretary for Water and Science at the Department of Interior from 1993 to 1995. Prior to 1993, she worked for the U.S. Geological Survey and the U.S. Senate Environment and Public Works Committee. She has a B.A. in chemistry from Wellesley College, an M.S. in civil engineering from the Massachusetts Institute of Technology, and a Ph.D. from the Department of Geography and Environmental Engineering at Johns Hopkins University. She is the 2001 recipient of the Johns Hopkins University Alumni Association's Woodrow Wilson Award for Distinguished Government Service.

Lawrence J. MacDonnell is an environmental and natural resources attorney with the firm of Porzak, Browning & Bushong in Boulder, Colorado. His practice emphasizes water law and the Endangered Species Act. He is active in several nonprofit organizations involved in community-based conservation in Colorado. Between 1983 and 1994, he served as the initial director of the Natural Resources Law Center at the University of Colorado School of Law, where he also taught courses in environmental and natural resources law. He has published and spoken widely. His most recent book is *From Reclamation to Sustainability: Water, Agriculture, and the Environment in the American West*. He is a past president of the Colorado Riparian Association. Dr. MacDonnell served on the NRC Committee on the Future of Irrigation, and the Committee on Riparian Zone Functioning and Strategies for Management. He received his B.A. from the University of Michigan, his J.D. from the University of Denver, and his Ph.D. from the Colorado School of Mines.

Thomas K. MacVicar is a private consultant specializing in the water resources of south Florida. He has been president of MacVicar, Federico & Lamb, Inc., since its inception in 1994. Prior to beginning his consulting practice, he spent 16 years on the staff of the South Florida Water Management District. From 1989 to 1994, he was second in command of the 1,500-employee agency with direct responsibility for all water resource issues in a 16-county jurisdiction from Orlando to Key West. His firm is currently involved in numerous complex water resource-related processes for both public and private sector clients. Mr. MacVicar participated in the multiyear process to develop the Comprehensive Everglades Restoration Plan on behalf of the Florida Department of Agriculture and Consumer Services and a broad cross section of agricultural landowners, businesses, and associations. He currently serves as chair of the Miami-Dade County Flood Management Task Force, created by the County Commission in response to back-to-back floods in 1999 and 2000. He received a B.A. in political science from the University of South Carolina, a B.S. in agricultural engineering from the University of Florida, and an M.S. in water resource engineering from Cornell University. Mr. MacVicar was a member of the NRC Committee on the Future of Irrigation in the Face of Competing Demands.

Rebecca T. Parkin is an associate professor in the Department of Environmental and Occupational Health with a joint appointment in the Department of Epidemiology and Biostatistics in the School of Public Health and Health Services at The George Washington University. She is also the associate dean for research and public health practice for the school and the scientific director of the Center for Risk Science and Public Health at the university. Previously, Dr. Parkin was director of Scientific, Professional and Section Affairs at the American Public Health Association and the assistant commissioner of the Division of Occupational and Environmental Health at the New Jersey Department of Health. Her areas of expertise include environmental epidemiology, public health policy, and risk assessment and communication. She has been a member of the NRC's Water

Science and Technology Board. Dr. Parkin received her A.B. in sociology from Cornell University; her Certificate in Science, Technology and Policy from Princeton University; and her M.P.H. in environmental health and Ph.D. in epidemiology from Yale University.

Roger K. Patterson is director of the Nebraska Department of Natural Resources. He was appointed by Governor Mike Johanns in January 1999. Prior to his appointment with the State of Nebraska, he spent 25 years with the Bureau of Reclamation, working in several of the western states. His last assignment with the Bureau of Reclamation was as regional director of the Mid-Pacific Region in Sacramento, California. Mr. Patterson has been honored with numerous awards throughout his career, among them the 1995 Presidential Rank Distinguished Executive Award, the highest award bestowed on a federal career employee. In 1999, Mr. Patterson received the DOI's Distinguished Service Award for achievements in addressing engineering, environmental, and public concerns as regional director of the Mid-Pacific Region. He is a graduate of the University of Nebraska with degrees in civil and sanitary engineering. In addition to being agency director, Mr. Patterson is chair of the Nebraska Boundary Compact Commission and is the state representative to the Missouri River Basin Association, the State Environmental Trust Board, the Blue River Compact, the Republican River Compact, and the Upper Niobrara River Compact.

Franklin W. Schwartz is a professor and the Ohio Eminent Scholar in hydrogeology at The Ohio State University. Dr. Schwartz's research interests encompass field and theoretical aspects of mass transport, contaminant hydrogeology, and watershed hydrology. He is coauthor of the texts *Physical and Chemical Hydrogeology*, published in 1990 and 1998, and *Foundations of Ground Water*, now in production. He has received various awards recognizing his contributions to hydrogeology, including the O. E. Meinzer Award, the Excellence in Science and Engineering Award, and the M. King Hubbert Science Award. He was elected as a fellow of the American Geophysical Union in 1992. In addition to his teaching and research, Dr. Schwartz acts as a consultant to government and industry, and he acts in various advisory capacities. He has served on various NRC panels and as a member of the Water Science and Technology Board. He received his Ph.D. in geology from the University of Illinois.

Amy K. Zander is a professor in the Department of Civil and Environmental Engineering at Clarkson University. She received a B.S. in biology and M.S. and Ph.D. degrees in civil engineering from the University of Minnesota. Her areas of expertise include drinking water treatment, treatment process design, membrane systems in environmental separations, life cycle assessment, and industrial ecology. Dr. Zander has received numerous awards for her research and teaching, including the 2003 Samuel Arnold Greeley Award from the American Society of

Civil Engineers for the paper that makes the most valuable contribution to the environmental engineering profession, the 2000 Association of Environmental Engineering and Science Professors/McGraw Hill award for Outstanding Teaching in Environmental Engineering and Science, and the 2001 Boeing Outstanding Educator Award. Prior to her academic career Dr. Zander was a water quality specialist with the Texas Water Commission, 1984–1986, and an engineer with James M. Montgomery Consulting Engineers in 1989. Dr. Zander served on the NRC Committee on Small Water Supply Systems.

Laura J. Ehlers is a senior staff officer for the Water Science and Technology Board of the National Research Council. Since joining the NRC in 1997, she has served as study director for nine committees, including the Committee to Review the New York City Watershed Management Strategy, the Committee on Riparian Zone Functioning and Strategies for Management, and the Committee on Bioavailability of Contaminants in Soils and Sediment. She received her B.S. from the California Institute of Technology, majoring in biology and engineering and applied science. She earned both an M.S.E. and a Ph.D. in environmental engineering at the Johns Hopkins University. Her dissertation, entitled RP4 Plasmid Transfer Among Strains of *Pseudomonas* in a Biofilm, was awarded the 1998 Parsons Engineering/Association of Environmental Engineering Professors award for best doctoral thesis.